ROUTLEDGE LIBRARY EDITIONS: GEOLOGY

Volume 24

THE NATURE OF GEOMORPHOLOGY

T0179066

THE NATURE OF
GEOMORPHOLOGY

ALISTAIR F. PITTY

Routledge
Taylor & Francis Group

LONDON AND NEW YORK

First published in 1982 by Methuen & Co Ltd

This edition first published in 2020
by Routledge
2 Park Square, Milton Park, Abingdon, Oxon OX14 4RN

and by Routledge
52 Vanderbilt Avenue, New York, NY 10017

Routledge is an imprint of the Taylor & Francis Group, an informa business

British Library Cataloguing in Publication Data
A catalogue record for this book is available from the British Library

ISBN: 978-0-367-18559-6 (Set)
ISBN: 978-0-429-19681-2 (Set) (ebk)
ISBN: 978-0-367-22087-7 (Volume 24) (hbk)
ISBN: 978-0-367-22415-8 (Volume 24) (pbk)
ISBN: 978-0-429-27475-6 (Volume 24) (ebk)

Publisher's Note
The publisher has gone to great lengths to ensure the quality of this reprint but
points out that some imperfections in the original copies may be apparent.

Disclaimer
The publisher has made every effort to trace copyright holders and would welcome
correspondence from those they have been unable to trace.

ALISTAIR F. PITTY

The Nature
of
Geomorphology

METHUEN LONDON AND NEW YORK

First published in 1982 by
Methuen & Co. Ltd
11 New Fetter Lane, London EC4P 4EE

Published in the USA by
Methuen & Co.
in association with Methuen, Inc.
733 Third Avenue, New York, NY 10017

Typeset by Keyset Composition, Colchester.
Printed in Great Britain
at the University Press, Cambridge

British Library Cataloguing
in Publication Data
Pitty, Alistair F.
The nature of geomorphology.
1. Geomorphology
I. Title
551.4   GB401.5

ISBN 0-416-32110-0
ISBN 0-416-32120-8 Pbk

Library of Congress Cataloging
in Publication Data
Pitty, Alistair F.
The nature of geomorphology.
Bibliography: p.
Includes index.
1. Geomorphology. 1. Title.
GB401.5.P67 1982   551.4 82-7921
ISBN 0-416-32110-0   AACR2
ISBN 0-416-32120-8 (pbk.)

Contents

Acknowledgements

The author and publishers would like to thank the following for permission to reproduce copyright figures:

The American Geographical Society for Fig. 12B; American Geophysical Union (and W. B. Langbein and S. A. Schumm) for Fig. 10D; *American Journal of Science* for Figs 2F (and A. Gupta), 8E (and B. Reed, C. J. Galvin and J. P. Miller) and 18A (and W. R. Farand); Association of American Geographers for Fig. 6A; *Australian Geographical Studies* for Figs 6B, 12F; The British Association for the Advancement of Science for Fig. 20B; *Bulletin of the Department of Geography, University of Tokyo* (and T. Yoshikawa) for Fig. 18B; Department of Mines and Technical Surveys, Ottawa, for Fig. 8D from *Geographical Bulletin*, reproduced with permission of Information Canada; *Die Erde* for Fig. 12E; Elsevier (and M. O. Hayes) for Fig. 15D; Elsevier (and A. Singer) for Figs 16A and 16B; *Erdkunde* for Fig. 15C; W. H. Freeman and Company for Fig. 8B; Gebrüder Borntraeger for Figs 5C, 8A, 14B, 15A; *Geografiska Annaler* for Figs 19B, 20E and G, 21B; The Geological Society of America for Figs 1B (and S. Uyeda and A. Miyashiro), 1D (and T. L. Henyey and T. C. Lee), 2A (and E. E. Larson *et al.*), 2B (and T. Matsuda, Y. Ota, M. Ando and N. Yokekura), 3 (and W. Alvarez), 4 (and V. R. Baker), 5B (and A. D. Howard), 7A (and K. L. Bowden and J. R. Wallis), 12A (and D. K. Davies, M. W. Quearry and S. B. Bonis), 13D (and J. M. Proffett), 15B (and T. J. Toy), 19C (and T. Sunamura), 20A (and R. A. Young and W. J. Brennan), 20C (and C. A. Bandoian and R. C.

Murray), 20D (and P. E. Calkin and C. E. Brett), 21A (and R. N. Dubois) and 21C (and M. G. Foley); The Geological Society of London (and W. S. Pitcher) for Fig. 1C; The Geological Society of London (and C. H. Scholz) for Fig. 2C; The Geologists Association for Fig. 14A; Figs 12C and D reproduced with slight modification from the *Journal of Glaciology* by permission of the International Glaciological Society (and J. F. Nye, E. R. LaChapelle and T. E. Lang); The Institute of British Geographers for Fig. 20F; Masson et Cie for Table 1; Methuen and Co. for Fig. 10B; Panstowowe Wydawnictwo Nankowe for Fig. 11 (from *Geographia Polonica*); *Quaternary Research* for Fig. 19A; The Royal Society (and R. W. Girdler) for Figs 13A, B and C; The Soil Science Society of America for Fig. 10C; The United States Geological Survey for Figs 17A and B; University of Cambridge Press for Fig. 10A, taken from *The Geological Magazine*; University of Chicago Press for the following figs taken from *Journal of Geology*: 1A, 2D and E, 7B and C, 8C, 16C and 22; Fig. 17C reprinted by permission of John Wiley and Sons Inc. (copyright © 1971).

As the present book represents a revision of the opening sections of *Introduction to Geomorphology*, first published in 1971, I remain indebted to all those mentioned in the Acknowledgements to the earlier work. I would also like to express enduring gratitude to Jay Appleton, as work on that book was initiated at his recommendation. Comments from Keith Clayton, Derek Mottershead and Nicholas Stephens on its presentation and format lent vital impetus to the task of revision. The completion of the present text and the draughting of its illustrations was made possible only by the generous support of Penny Ann Pitty. The scale of the problems encountered was considerably cut down to size by cheerful dialogues between Edward and Alice Pitty on how they might answer their hypothetical examination question: 'Describe the following glacial features – eskers, drumlins, hummocky moraine, striations, and kame terraces. Rank in order of increasing interest'.

It is hoped that the text reflects a little of the excellence of reading facilities enjoyed in the Brynmor Jones Library of the University of Hull, and the professional commitment of its staff. The text should also reflect the sustained and detailed editorial attention by the publishers, for which I am most grateful.

ALISTAIR F. PITTY
Cottingham
East Yorkshire

Preface

Geomorphological literature has now become so voluminous that it may seem impractical for an individual writer to attempt to encapsulate it all. Indeed, in the last ten years, the geomorphological market stall has been increasingly stocked with a range of specialized hybrids and a wealth of produce from academic communes involving several workers. However, if the individual student is to be expected to comprehend the scope of geomorphology, clearly it is only fair to expect that a sufficient overview must be within the grasp of the individual writer. Thus, the very reason that cautions against the present exercise is equally its main justification. Ideally, the advantage could be greater cohesion in scope, balance in format and evenness in level of treatment. Therefore, since each of these three aspects stems from different sources in the present blend, some brief introduction to its scope, format and level of treatment is required.

The scope of geomorphology envisaged in this book partly reflects the writer's earliest impressions of geomorphological phenomena. These include vivid childhood experiences of avalanching sand dunes, surging moorland streams, a footpath trenched overnight by snowmelt, and the endless fascination of diverting a straightened brook into an adjacent, arc-shaped hollow. No cyclist can fail to be aware of landform, which for me became a physical and emotional experience, with the spacing of contours keenly felt on climbs, and the risks in descent closely linked to accuracy in assessments of contour curvature. Some readers may therefore find that the scope envisaged here for geomorphology does not coincide with the formalities of some broadly established school of thought or text with which they are already familiar.

The format of the book perhaps leans too heavily on the human frailty of polarizing issues into two apparently opposed points of view. Such a duality in the opening section of the book, dealing with the scientific context of geomorphology, is inevitable. This is due to the duality established by geomorphology being largely part of geography in Europe, but remaining more commonly a component of geology in the United States. In the central section of the book, the nature of geomorphology is also examined under the headings of the opposed fringes of spectra of opinion, such as 'process and form' or 'laboratory and field work'. For the basic postulates of geomorphology discussed in the concluding section, dualities are again featured, as between 'stillstands' and 'the mobility of earth structures'. However, the format equally depends on ranging some of the arguments for and against certain concepts, like the 'Cycle of Erosion' or that of 'morphoclimatic zones'. In all cases where a duality is taken as a convenient if arbitrary starting point, particular attention is paid to identifying any broad area of common ground between extremes of emphasis. Indeed, the attention given to describing the extreme poles of viewpoints may sometimes appear too scant. Another emphasis in the format is the identification of those third and even fourth directions that are often overlooked when discussions only in terms of alternatives prevail.

A consistent level of treatment is not easily maintained for geomorphological subject matter, partly due to a unique aspect of geomorphology amongst the components of physical geography. As initially presented to the English-speaking world, physical geography was 'a description of the earth, the sea, and the air, with their inhabitants animal and vegetable, of the distribution of these organized beings, and the causes of that distribution' (M. Somerville, *Physical Geography*, 1848). Subsequently, each of these components, although intrinsically interrelated, has been investigated and formalized in increasingly separate ways. Their subject matter, therefore, is not incorporated into advanced training in geography in the same style. Geology and ecology can be taught by experts from those academic departments, often neighbours on the same campus, or botanists may be recruited to work within a geography department. The experts in meteorology, hydrology and soil sciences, however, are largely located outside educational institutions. Therefore, physical geographers usually have to translate the specialist findings of these technologies into appropriate teaching materials. Clear examples of this undertaking include the *Climatology* of A. A. Miller (1931) and the *Principles of Hydrology* by R. C. Ward (1967). Some geographers may still acquire their geomorphology from geologists during training. More commonly, however, those responsible for the geomorphological component in a geographer's training have neither other academic departments from which their students might benefit nor specialized institutions providing a coherent bulk of knowledge. A distinctive feature of geomorphology, therefore, is the degree to which geographers provide the bulk of the necessary knowledge themselves. Set against the obvious advantages of such a continuity over a wide range of responsibilities is the absence of arbitrary aids in distinguishing between stimulating research and relevant teaching materials in the making. In consequence, it is not always easy to write on geomorphology consistently at the same level of specialism. This unique aspect of geomorphology amongst the other components of physical geography may also explain why geomorphology is often singled out for scrutiny when attempts to circumscribe geography become oppressive.

Scientific context of geomorphology 1

First, then, as to the configuration of the earth's surface. We have here a bone of contention between the geographers and the geologists. The latter hold that the causes which have determined the form of the lithosphere are dealt with by their science, and that there is neither room nor necessity for the physical geographer. (Mackinder, 1887)

Geomorphology itself has suffered, and will continue to suffer, from attempts to include it in the geographic realm. In the history of its development, in its methods, and in its affiliations it is a part of geology. (Johnson, 1929)

The place in geography of geomorphology is still a debated question, and though it enjoys a privileged place abroad, its claim to inclusion within the main terrain of study is commonly denied by geographers of the 'inner faith' in this country. (Wooldridge, 1932)

Introduction

Wooldridge (1958) regarded it 'as quite fundamental that geomorphology is primarily concerned with the interpretation of forms, not the study of processes. The latter can be left to physical geology'. On the other hand, 'It would not be possible to predict the path that a drop of water falling on the slope would take. I cannot imagine a geologist wanting to do so' (Kitts, 1976). Despite reiteration of such disciplinary guidelines, however, some contributions from geography concentrate on non-geographical aspects of geomorphology and others, professionally designated as geologists, are leading practitioners of what Kitts finds unimaginable. Although the 1960s saw much reshaping of geomorphology, the question of where it belongs in research, applied study and practice, or in advanced teaching, was unresolved (Dury, 1972). Thus today the scientific study of landforms and landforming processes comprises a range of differing understandings of its purpose, context and scope. Regional emphases are easily recognized and appreciated, but disciplinary shades merge or clash whilst institutional styles or personality tinges may glare or pass unnoticed.

Geomorphologists are academic migrants, a vociferous, squabbling yet cohesive flock, tracking continually across the political boundaries of the established nation-states of science. Roosting comfortably in some geological branches, they are the ugly ducklings in other nests, and some uneasy clutches of geographers suspect that the raising of a boisterous cuckoo has been foisted on them (Brown, 1975). Thus, Gould (1973) reflects that 'the excising of physical geography is not nearly so traumatic as many would have us believe', whereas Chorley (1971) recommends that physical geography must be sensitive to the changing aims and objectives of human geography and the allied humanities. It is, therefore, important to identify points at which geomorphologists' interests merge or contrast with the main contemporary purposes of either geology or geography, not least because both have changed substantially during the 1970s.

Implications of recent changes of emphasis in geology

PLATE TECTONICS

Since 1970, when the term 'plate tectonics' began to appear in print, geologists have restructured much of their knowledge of geophysical processes within this new, developing conceptual framework. In contrast to the early scepticism which first greeted A. Wegener's ideas on continental drift, geologists now generally accept that there is 'a continuous network of mobile belts about the earth which divide the surface into several large rigid plates' (Wilson, 1965). For geomorphology, the concept of lithospheric plates is probably most important in outlining areas of relative vertical stability and zones of intense deformation (Rice, 1977; 76). These rigid plates, which move on top of a mechanically soft layer, the asthenosphere, have three types of boundary that depend on directions of plate motion. First, trailing edges are identified where plates move away from each other, with material being added from great depth, mostly along mid-oceanic ridges. Secondly, where plates collide or converge, the leading edge boundary is identified, with an ocean trench marking the zone where the overridden plate begins to descend the subduction zone which leads down to the asthenosphere (Fig. 1). Where the converging plates carry no continents, the main morphological expression of plate boundaries is ocean trenches and island arcs, as in the Pacific. If the leading edge of such an oceanic plate is underthrust below a continental plate, high mountain ranges like the Rockies and Andes are created and maintained. The

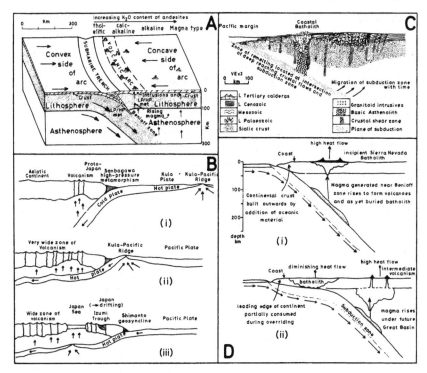

Figure 1 Geographical contrasts and similarities in major earth-surface features explained by subduction hypotheses

A   Some features of island arcs at the convergence of two lithospheric plates. LP/HT indicates low-pressure/high-temperature metamorphism.

B   Tectonic history of Japan; related to collision and descent of the formerly mid-oceanic Kula-Pacific Ridge in Late Cretaceous times.

(i) 120 million years BP. Ridge approaches Asiatic continent.

(ii) 90 million years BP. Thermal effect of underthrust plate causes extensive vulcanism.

(iii) 70 million years BP. Thermal effect further reduces thickness of part of the continental plate, which is eventually broken by tensional force, the fragment drifting away to form the Japanese islands.

C   Possible generation of the Coastal Batholith of Peru in the Lower Tertiary along deep-penetrating faults intersecting east-migrating subduction zone.

D   Tectonic and geothermal history of the Sierra Nevada and the initiation of the Basin-and-Range province.

(i) Before 80 million–100 million years BP. Possible generation of the Sierra Nevada batholith from underthrusting slab in Mesozoic times.

(ii) 80 million–100 million years BP to 20 million–30 million years BP. Cold lithospheric slab beneath Sierra crust with steeply inclined portion of subduction zone and associated magma generation having shifted east, attributable to increased westerly drift of the North American plate.

Sources: Adapted from A Mitchell and Reading, 1971; B Uyeda and Miyashiro, 1974; C Pitcher, 1978; D Henyey and Lee, 1976.

Himalayas owe their greater altitude to the collision of two continental plates. A third type of plate boundary may be marked by deep fractures along which plates slip past each other laterally, with little gain or loss of surface material. In such cases, like California, surface strike-slip faulting has occurred during historically recorded earthquakes and major faults have moved during the Quaternary. In contrast, in some continental interiors, stable 'cratons' of old, resistant rocks may have remained largely unmoved since Precambrian times.

Clearly, the plate-tectonic model illuminates the distinctive character of many of the earth's major surface configurations. Some of these are reflected in larger-scale geomorphological responses, such as continental asymmetry, with high orogenic belts on the collision side of continents explaining the general absence of large rivers on the 'leading-edge' sides of North and South America (Inman and Nordstrom, 1971). Conversely, trailing-edge continental coasts receive more sediment than collision coasts because of the larger drainage areas (Potter, 1978), and a further geomorphological consequence is the correlation of percentage of coast-line length occupied by barrier islands with plate-margin type (Glaeser, 1978). Where strike-slip faulting is active, deposits and landsurfaces broken and offset by these displacements can be surveyed and mapped, with numerous examples from New Zealand already enshrined in the works of Cotton (1960). Not least, the combined effect of continental drift and the wandering of the earth's poles may have initiated the late Cenozoic glaciations when the South Pole and the drifting antarctic continent began to overlap.

DEEP-SEA EXPLORATION

Since 1968, when the vessel *Glomar Challenger* began deep-sea drilling, a worldwide network of ocean-floor cores has been established (Wyllie, 1976). This Deep Sea Drilling Project, advised by the Joint Oceano-graphic Institution for Deep Earth Sampling (JOIDES) has supplemented and extended the intensive explorations of the oil industry which first moved offshore in the 1940s. Thus, the evidence for plate tectonics and the oppor-tunities for the oil industry have led the emphasis of geology away from the dry land and its forms, into offshore, continental shelf and deep-sea areas. How-ever, evidence indirectly relevant to geomorphology has been obtained, simply because accumulation rates and stratigraphy of offshore sedimentation are the natural complement to the erosional record on land. For instance, the scale of the Navy Fan, offshore from the Tijuana River and adjacent streams draining into San Diego Bay, suggests that the mean rate of late-Quaternary denudation of northernmost Baja California is 60 cm/1000 yr (Normark and Piper, 1972). Further south, in enclosed basins on the Pacific continental margin of Central and South America, a median sedimentation rate of 8.8 cm/1000 yr has been recorded.

Episodes of tectonic activity, vulcanism and climatic change can be inferred from cores. For example, the 800 × 300 km Meiji sediment tongue occurs in the north-west corner of the Pacific floor. Because no major Siberian river could be involved, this 1800 m thick sedimentary body may date the initiation of the steep coastal drainage system and the associated uplift phase of an early Upper Tertiary orogenic pulse (Scholl *et al.*, 1977). Recurrence and intensity of volcanic activity is identifiable in some cores, particularly for basin-wide episodes of stagnation in the Mediterranean Sea. More than forty tephra layers occur in deeper horizons of the Tyrrhenian Sea and, by matching the mineralogy and petrochemistry of deep-sea ashes with their source regions, a time scale for many of the major Mediterranean eruptions has been established (Keller *et al.*, 1978). The influence of climatic fluctuations can be estimated from oceanic and continental responses because both pelagic and terrigenous components are recognizable in deep-sea cores. The distribution and accumulation rates of clay and terrigenous silts and sands indicate continental responses to climatic fluctuations. Glacial–interglacial change in the temperature and salinity of the ocean is reflected by planktonic foraminifera and, in lower latitudes, by amounts of pelagic carbonate. In higher latitudes, glacial intervals can be identified by increases in the amount of ice-rafted detritus in cores and by the occurrence of erratics beyond the present-day limit of icebergs. Core analyses from JOIDES leg 28 in the Ross Sea suggest a slow rate of sedimentation in the Miocene up to about 5 million years BP, followed by rapid sedimentation of ice-rafted debris at the Mio-Pliocene boundary. The sequence suggests a glacial maximum followed by melting, collapse and retreat of the ice sheet and implies extensive erosion in the Transantarctic Mountain source area (Drewry, 1975).

ENGINEERING AND ENVIRONMENTAL GEOLOGY

Significantly, the founder of modern geology, William Smith, was a canal engineer, and the escalation of urbanization and the extension of associated road networks in recent decades is reflected in renewed attention to engineering geology. This practice involves the general principles of geology as they may influence constructional project designs, particularly the physical properties of rock stability and bearing strength (Woodland, 1968). The variation in lithology and the structural relations of rocks beneath the ground-surface must be fully appreciated in the appropriate siting and development of major civil-engineering projects. For example, the layout of the engineering works of the Cruachan Pumped-Storage Scheme followed closely geological mapping which delineated uniform conditions within a subsidence pluton of granodiorite. Geology is also a major influence on motorway construction costs, as these are dependent on the character and scale of earthworks. Geological influences are partly expressed in the configuration of the land-surface over which routes might be planned. These influences are accentuated where landsurface conditions vary significantly along the route, modifying the suitability of materials for embankment construction, the declivity for safe side slopes for embankments and cuttings and the volume of rock to be excavated (Newbery and Subramaniam, 1977). The renewed growth of engineering geology increases knowledge about rock mechanics, a fundamental geomorphological property. Equally important is the attention now paid to 'soft' rocks, since the youngest drift deposits are commonly a major consideration in siting civil engineering projects, with hidden, drift-filled valleys a particular problem in reservoir construction. Not least, many engineering failures are found to be related to instability of hillslopes. For example, a whole section of a partially completed bypass road near Sevenoaks, Kent, was abandoned. Construction reactivated movement in extensive soli-fluction deposits and in landslips in the Hythe Beds and the underlying Atherfield and Weald clays. It was revealed that several solifluction lobes were superimposed, each with a marked slip-plane of low shear strength at its base (Woodland, 1968).

Environmental geology is based on hydrogeology and branches of economic geology as well as incorporating engineering geology. It is linked with other earth-science specialisms, including marine geology, sedimentology, seis-mology and geomorphology, thus embracing 'those aspects of geology which are concerned with the use, by society, of the Earth and its resources and, in addition, the implications of geological processes on man' (Knill, 1970). In addition to the urban planning and site-foundation responsibilities of engineer-ing geology, environmental geology considers natural hazards, seen at their most sensational in earthquakes and volcanic eruptions. For example, it is possible to plot on a relief map the area in which an eruption, should it occur, would produce flows long enough to encounter a town (Guest and Murray, 1979). Man-made hazards are also considered, with the location of sites where waste products can be safely dumped becoming an increasingly exacting task.

Since 1970 environmental geology has been incorporated into the curriculum of an increasing number of colleges and universities, attracting significantly increased student enrolment (Hoffman, 1979). In addition, a model for a humanistically orientated undergraduate geology course has been outlined (Romey, 1972).

Divergence between geomorphology and certain emphases in geology

For the geomorphologist, the receding sea-cliff is a striking erosional land-form; to the geologist, it is a fresh exposure of rocks (Brown and Waters, 1974). Similarly, the geomorphological weathering processes that have fashioned and continue to modify landforms are, for the geologist, essentially 'subaerial diagenesis', the processes by which new rocks are being formed, with com-parable products identifiable in the stratigraphic record. Differences in per-ception of the same phenomena are common. Perhaps the clearest instance of divergence is seen in the intense interest of geologists in vulcanicity. Admittedly, geomorphologically significant volumes of lava and clastic material can be derived from active volcanoes and the geographical spread of ash can be on intracontinental scales. However, geologically crucial glimpses are offered into the largely inscrutable characteristics of magma generation in the earth's molten interior and guides to the unaltered composition of the magmatic melt phase. Thus, ash sheets, uncommon and largely expressionless as landforms, are exceptionally interesting to geologists. Where ash composition is zoned, it reflects, in inverted order, the original compositional zonation within their magma-chamber source. Study of Quaternary volcanic domes and intrusive rocks also provides vital insights into the processes of magma generation, but these are either small or geographically uncommon landsurface features.

There are instances where geological enquiry ceases to be directly concerned with the study of shape of the present landsurface. These arise because geology, being an historical science, is concerned with occurrences at particular times and places in the past. However, due to the time-span during which the present landsurface has been taking shape, geomorphologists are readily drawn into purely stratigraphical questions, particularly those of Quaternary chronology (Clayton, 1980). For instance the sequence of events in the Pleistocene succession of the Irish Sea can be based exclusively on the stratigraphical record (Bowen, 1973). Even where some stratigraphical units are at the landsurface, like the Cheltenham Sand and Gravel, they may have no landform expression (Briggs, 1975). In the case of East Anglia, where Pleistocene deposits provide the most complete stratigraphical record for this period in Britain, geomorphological studies of glaciofluvial valley trains, terraces and tills are rare (Straw, 1973). Certainly, some striking geomorphological suggestions may be briefly made, such as the height range of the Beestonian sands and gravels in southern East Anglia possibly representing a series of river terraces found at progressively lower levels as a much larger proto-Thames drainage system migrated south-east (Rose, Allen and Hey, 1976). Also, even the stratigraphy alone, in some featureless coastal locations, may reflect a history of change in plan shape. For instance, on the coastal fringe of the Somerset Levels, the presence of a buried shingle ridge below the present storm beach at Stolford demonstrates that the coastline has remained stable over the last 6000 years (Kidson and Heyworth, 1976). In general, however, the geomorphological interpretations extracted from Quaternary stratigraphy may be minimal (Green and McGregor, 1980). Even where river terraces are dominant landforms, little may have been done to establish the nature of river activity at various stages in the Pleistocene or to relate the pattern of river activity to the stratigraphical evidence.

These examples are drawn from one area only since the stratigraphy and paleontology of every landsurface of glacial deposition is complex, varied and generally so localized that broader patterns are not always recognizable. In most of these areas, therefore, a local terminology and classification is preferred, such as those which describe the striking oscillations between glacial and interglacial conditions in North America in the latter part of the Pleistocene (Wright and Frey, 1965). In the Alpine valleys of Europe, the terms for the deposits of a glacial sequence – Günz, Mindel, Riss and Würm – have been utilized for more than a century.

There are several examples of landforms being studied very intensively as aids in elucidating the last phases of the geological record whilst their geographical scale and occurrence is limited. For example, the responses of coastal barrier islands to Pleistocene fluctuations of sealevel has been much debated in order to devise satisfactory hypotheses of the processes and effects of marine transgressions and regressions. Similarly, on tectonically rising coasts, wave-cut notches become the focal point of the stratigrapher's interest. Conversely, on sinking coasts, it would be too arbitrary to draw a deciding limit to *landforms* at the present stand of sealevel. For instance, offshore from southern New England an 18 km-long continuation of the Old Saybrook moraine can be recognized. Elsewhere, some wave-cut notches as well as drowned valleys are geomorphologically significant as relict landforms. However, when submarine slumps, slide sheets and similar mass movements are described from the modern marine environment in order to comprehend characteristics of ancient sedimentary rocks, the specifically geological purpose is evident.

Geographical characteristics of geomorphology

LANDFORM GEOLOGY

Many geological phenomena that influence landsurface forms have the intrinsically geographical characteristics of spatial pattern and regional distinctiveness. At the broadest scale these are now more clearly illuminated by the plate-tectonic model (see p. 2). For instance, unlike the older geosynclinal hypothesis, plate tectonics does not require a regular sequence of events for orogenesis in all mountain belts. Thus, the tectonic style of the Andes differs from that of the Alps in that folding is upright and nappes are absent. Conversely, the western High Atlas and the Jura are very similar in structural and geomorphological features although the tectonic settings are quite different. In the case of vertical movements of continents as a whole, Africa is unusual in that its surface has risen significantly in relation to other continents (Bond, 1978).

Most major tectonic activity on Earth is strictly localized along plate boundaries, as the cold, seismically active subducted plates descend into the asthenosphere. Earthquake occurrence is therefore very geographically circumscribed, with more than 90 per cent of intermediate-focus earthquakes and nearly all of the deep-focus earthquakes occurring in the circum-Pacific belt (Fig. 2A and B). Even the curved form of island arcs is explained, representing the trace on a spherical earth-surface of the downbending of a slab of lithosphere as it enters the subduction zone. Due to the melting of the down-going slab beyond a certain depth, this same belt is also the 'ring of fire', incorporating some two-thirds of the world's active volcanoes. Strike-slip faults may vary in seismic style, but instructive comparisons can be made between the San Andreas fault of California and the Alpine Fault of New Zealand (Fig. 2C). Volcanic forms may also vary laterally (Fig. 2D) and the finer products of volcanic eruptions, in turn, form geographical patterns downwind as they are continually modified into several discrete sub-populations of particles during upward, outward and downward movement from vents (Fig. 2E). Indeed, there are many ways in which vulcanicity demonstrates a fundamentally geographical element which may be present in geological phenomena (Clapperton, 1977).

GEOGRAPHICAL CHARACTER OF GEOMORPHOLOGICAL PROCESSES AND FORMS

Geomorphological variability is also often intrinsically spatial in nature. On the broadest scale, contrasts between hemispheres can be noted. Thus, unlike many regions of the northern hemisphere where the glacial chronology is the key to much of Quaternary history, the Australian glacial ice accumulations were only a minor feature of that period (Peterson, 1968). At lesser scales, erosional forms, like pediments, have been noted to vary from region to region. In some river valleys, broad contrasts are noted, as between rocky upland stream channels and more mobile, smoother beds of lowland rivers. In the ephemeral stream channels of the American south-west, erosional and depositional environments are geographically separated, with the headcut incision contrasted with increased rates of deposition downstream from the knickpoint (Patton and Schumm, 1981). Often landslides are located where local oversteepening has occurred (Fig. 2F).

Aspect and orientation are important, as on sheltered coasts which are low-energy environments with minimal wave action compared with the heavy swell of more open coasts. Quite often changes follow systematically in a given direction, like downstream increases in channel width. Also, due to the inescapable influence of gravity, behaviour of weathered debris is intrinsically geographical. Even if the short-term movement of a particle is random, there is inevitably a long-term drift, such as downslope, downstream or alongshore, with associated trends or 'clines' in morphological expressions or sedimentary structures and forms. Glacial deposits, localized along major ice-marginal positions, may have marked longitudinal trends, particularly where basal sliding was important, as demonstrated in the Prairies of North America (Moran *et al.*, 1980).

In some cases, so many local influences are significant that the total expression of process and form in that locality is highly distinctive. Thus, equations for summarizing discharge data in one area rarely match measurements made in other areas, and at river mouths depositional forms and sedimentary sequences are the most varied of all coastal accumulation forms. Whether to establish broad similarities or to encompass local variations, a geographically varied and large number of instances may have to be considered if meaningful geomorphological generalizations are attempted. For instance, in studying chemical denudation at twelve sites in Devon, Walling and Webb (1978) observed a range between 6 and 29 m3/km²/yr for the solute load. A geographical variability of such a range within adjacent subdivisions of a small region emphasizes the risks of generalizing about erosional processes from a widely scattered series of single cases.

SCALE

Geographical scale is a fundamental characteristic of landforms, and a broad framework of size orders has been suggested (Table 1). Differing size orders may be more appropriate for specific groups of landforms. Coastlines illustrate the effect of scale particularly clearly, with the largest-scale views revealing much parallelism with known tectonic features. However, the degree of association decreases as the size of the features diminishes. When configurations are linked with moving tectonic plates, major-order features are about 1000 km along coasts, the on-offshore dimensions of about 100 km include continental shelves and coastal plains, and the height range from ocean floor to summits of

---

Figure 2 Some geographical characteristics of active portions of the earth's surface

A   Distribution and curvature of island arcs, convex towards ocean basins.

B   Distribution and recurrence of major earthquakes in central Honshu. Earthquakes are due to faulting or thrusting between oceanic plates and the Asian continental plate. Source areas are inferred from tsunami refraction patterns. The dashed line indicates an area predicted as the focus for a future earthquake.

C   Regional comparison of seismic history and pattern in major tectonic elements along transform (strike-slip) faults which transect continental lithosphere. Contrasts include uplift which is produced by thrusting on the Alpine fault but is distributed among the Transverse Ranges faults in California.

D   Geographical trend in altitude of composite cones in the Quaternary Central American volcanic chain. The trend is due to geographical variation in density of primary magma and a limiting critical altitude above which lava cannot be erupted, cones being at lowest altitudes where densities are highest.

E   Geographical trend in size and sorting of volcanic ash deposits in Iceland with increased distance from the vent. Size and sorting are expressed in *phi* units (see Pitty, 1971); 130).

F   Distribution of landslides in two basins in eastern Jamaica, reflecting the importance of stream-bank undercutting (Gupta, 1975). The similar density in pattern is established despite contrasts in amount and intensity of seasonal precipitation in the two basins.

Sources: Adapted from A Larson *et al.*, 1975; B Matsuda *et al.*, 1978; C Scholz, 1977; D Rose *et al.*, 1977, based on data in Mooser, Meyer-Abich and McBirney, 1958; E Walker, 1971; F Gupta, 1975.

coastal mountains is some 10 km (Inman and Nordstrom, 1971). Erosion and deposition may modify major-order features to those of intermediate sizes, approximately 100 km long, 10 km wide and 1 km in height. Wave action accounts for smaller-order features, such as beach face and berm or longshore bars. These also depend on the type of material deposited and the sedimentary structures associated with the nearshore transporting and depositional processes. These features are commonly 1–100 km alongshore and 10 m–1 km in the on-offshore dimensions. In general, most geomorphological enquiries concentrate at scales where significant subaerial modification of tectonic lineaments is identifiable.

Some landforms may vary in size, whereas others may not. In the Canadian Cordillera, for instance, larger cirques tend to be well below the snowline, whereas the smallest are well above (Trenhaile, 1976). Although of widely ranging scales, river meanders have the same ratio of dimensions in plan, regardless of their size. One of the most commonly reported indices of relief, Schumm's relief ratio, tends to vary inversely with drainage basin magnitude (Abrahams, 1977). Basin area affects sediment yield, with lower yields from a

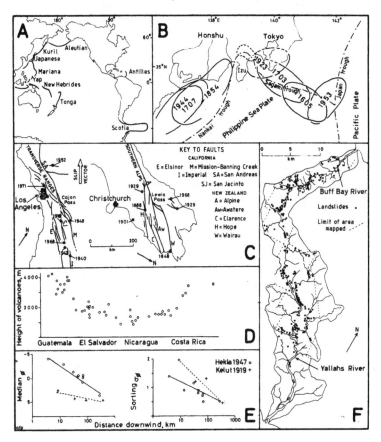

Table 1   Classification of geomorphological features (after Tricart, 1965)

Order	Units of earth's surface in km^2	Characteristics of units, with examples	Equivalent climatic units	Basic mechanisms controlling the relief	Time-span of persistence
I	10^7	continents, ocean basins	large zonal systems controlled by astronomical factors	differentiation of earth's crust between sial and sima	10^9 years
II	10^6	large structural entities (Scandinavian Shield, Tethys, Congo basin)	broad climatic types (influence of geographical factors on astronomical factors)	crustal movements, as in the formation of geosynclines. Climatic influence on dissection	10^8 years
III	10^4	main structural units (Paris basin, Jura, Massif Central)	subdivisions of the broad climatic types, but with little significance for erosion	tectonic units having a link with paleogeography; erosion rates influenced by lithology	10^7 years
IV	10^2	basic tectonic units; mountain massifs, horsts, fault troughs	regional climates influenced predominantly by geographical factors, especially in mountainous areas	influenced predominantly by tectonic factors; secondarily by lithology	10^7 years
			limit of isostatic adjustments		
V	10	tectonic irregularities, anticlines, synclines, hills, valleys	local climate, influenced by pattern of relief; adret, ubac, altitudinal effects	predominance of lithology and static aspects of structure	10^6–10^7 years
VI	10^{-2}	landforms; ridges, terraces, cirques, moraines, debris, etc.	mesoclimate, directly linked to the landform, e.g. nivation hollow	predominance of processes, influenced by lithology	10^4 years
VII	10^{-6}	microforms; solifluction lobes, polygonal soils, nebka, badland gullies	microclimate, directly linked with the form, e.g. lapies (karren)	predominance of processes, influenced by lithology	10^2 years
VIII	10^{-8}	microscopic, e.g. details of solution and polishing	micro-environment	related to processes and to rock texture	

unit area in a larger basin than from the same area in a small basin. In general, sediment discharge/km^2 of basin area decreases by about a factor of two for each ten-fold increase in total basin area (Wilson, 1973). In eolian environments, the coarsest material collects on the crests of small-scale sand forms, whereas the reverse obtains for large-scale dunes.

Scale influences methods used to collect data and indicates which specialism, from geophysics to sedimentology and soil science, will illuminate the significance of the results. For instance, geophysicists do not consider plate tectonics at scales of a few tens of kilometres, especially in complex areas where 'microplates', if identifiable, are not strictly rigid. This perhaps sets an upper limit to the size of strictly geomorphological forms. At the other extreme, the common tendency to discount particles larger than 2 mm in soil science and clasts larger than pebbles in sedimentology, indicates the lowest limits at which microrelief might be recognized. The scale most commonly considered by the geomorphologist is the measuring interval selected for hillslope survey (Gerrard and Robinson, 1971).

The scale of geomorphological enquiry is also linked to the cost of research. For instance, to provide the critical data on the details of the sea-floor spreading process at the Mid-Atlantic Ridge, the Project FAMOUS involved numerous cruises in manned submersibles and major projects for outfitting and training; the US Geological Survey can instrument the entire Mississippi. However, for research with which the individual student can directly identify and perhaps incorporate into his or her own dissertation work, tape-measure widths and rubber-boot depths are the most relevant.

Implications of changes of emphasis in geography

We all feel . . . that somewhere in geography there is a fundamental unity which
eludes us. Is not our difficulty how to weld together the geological and the
human aspects of the subject? . . . Is it not perhaps the lure of geomorphology
which has been misleading us? (Mackinder, 1931)

INTRODUCTION

One early answer to Mackinder's query was perhaps the title *The Physical Basis of Geography: An Outline of Geomorphology* (Wooldridge and Morgan, 1937). Today, at the risk of perpetuating seemingly interminable and, at times, acrimonious discussion, the nature, implications and repercussions of geomorphology's relations with geography require continuing examination – for two reasons. First, in the light of accumulating knowledge, any clarification of geomorphology's dual nationality might make the distinctiveness of this specialism more readily appreciated. Secondly, internal priorities and external influences continue to change, so indelible or ephemeral features of the interrelationship between geography and geomorphology may become gradually more discernible. Such re-examinations assist the student to assess whether the emphasis on geomorphology in either geography or geology is appropriately weighted.

THE 1930S: ENVIRONMENTAL EMPHASES IN GEOGRAPHY

One of the most striking changes in the relations between geography and geomorphology occurred in the 1920s. This was the widely observed 'remarkable decline' in geomorphological contributions to American geography noted by Johnson (1929), who also recommended the exclusion of geomorphology as a first step in delimiting the field of geography. Admittedly, Leighly (1955) reflected that the motivation for the restricted redefinition of geography as 'human ecology' was obscure, but certain possibilities are discernible. Previously, the founder of the Association of American Geographers had decreed that 'any statement is of geographic quality if it contains . . . some relation between an element of inorganic control and one of organic response' (Davis, 1906). This assertion conforms with the interpretations of 'environmental determinism' that were then in vogue, an extreme view of the age-old questioning of the influence of the physical environment on mankind, in which human beings were regarded as the product of the earth's surface. In consequence, *causal connections* between physical and cultural phenomena were greatly stressed until 1930, when many geographers became increasingly eager to avoid references to the physical environment, in order to dissociate themselves from the 'environmental determinists' ' view as it became discredited. Apart from fingers burnt and the feet frozen as the environmental determinism collapsed around the ears of the followers of Huntington and Semple, pragmatic considerations were also persuasive. Geographers deliberately sought interests that geologists could not claim, partly in order to gain administrative independence (Sauer, 1941). This possibility is corroborated by experience in Britain, where 'geologists had captured physical geography, and had laid it out as a garden for themselves' (Mackinder, 1921). More generally, the 'uneasy feeling grew that the presence or absence of an environmental parameter in a study was not an operationally effective way of distinguishing geography from other disciplines' (Taaffe, 1974).

THE 1940S AND 1950S: HARTSHORNE'S VIEW OF GEOGRAPHY

A conspicuous landmark in the methodology of geography for English-speaking students is the work of Hartshorne, with the value of his *The Nature of Geography* (1939) being reinforced by the appearance of his *Perspectives of the Nature of Geography* twenty years later. The time-span covered is, in effect, much greater since Hartshorne depended heavily on translating Alfred Hettner who, in turn, had provided a summary of geography as it had evolved in Germany in the nineteenth century. The reason why Hartshorne's first statement contains barely any reference to geomorphology might therefore be most readily sought in Phillip Tilley's translation of the second, 1928 edition of Hettner's *Die Oberflachenformen des Festlandes* (Hettner, 1972). This work, if not a tirade against Davis, is a tenacious rejection of Davisian geomorphology, since 'his scheme lacks vitality, the landscape picture it gives has a moribund and dismal emptiness' (Hettner, 1972; 135). In consequence, it is F. F. von Richtofen who features in Hartshorne's translations as 'the founder of modern geomorphology'.

On reflection and, as several comments and citations suggest, under considerable challenge from Wooldridge, Hartshorne (1959) gave geomorphology much greater attention. However, he found it 'somewhat disconcerting that insistence on the value of such analyses for geography comes primarily from those who produce them rather than from those who use them'. The converse, that of social scientists prescribing priorities for geomorphology, has more far-reaching consequences where geographers are engaged in sustaining geomorphology as a distinctive natural science.

THE 1960S: SOCIAL SCIENCE AND STATISTICS IN GEOGRAPHY

The geographic profession now includes a substantial, if not dominant, group
of scholars who hope to see their subject ranked alongside anthropology,
economics, political science, psychology and sociology as a social or behavioral
science. (Mikesell, 1969)

By the mid-1960s, American geography was very different from that at the beginning of the 1950s, and by the late 1960s the change had spread to Britain, Canada and Australia (Taaffe, 1979). If concern about the place of physical geography had once seemed to be a specifically American problem, this has become increasingly widespread. Inevitably, as waves of urbanization advanced, a correspondingly greater emphasis on urban geography and social forces washed over geography, entraining a more abstract and theoretical approach. In particular, early quantification and theoretical work was drawn to the relative simplicity and geometrical regularity of transportation networks, precisely condensed into statistical and mathematical form. Law-seeking, 'nomothetic' generalizations were preferred to 'idiographic' approaches, emphasizing the individual and the unique. Thus, the regional geography reviewed by Hartshorne and the poetic, rural regions of Vidal de la Blache were brushed aside by the symbols of statistical and mathematical models representing the spatial properties of urbanization. Students of geography themselves came increasingly from urban backgrounds and were intrigued by quantitative geography's contributions to planning, whether within a *laissez-faire* or state-centralized political setting. Not least, the more explicit emphasis on spatial analysis reinforced an anthropocentric view of geography, simply because it is man's distinctive trademark to organize space geometrically. From the Central Business District to the moorland wall, snowline, or desert fringe, regularity of arrangement and functional relationships between widely separated places makes spatial analysis inevitably a man-centred if not an urbanocentric viewpoint (Pitty, 1979). 'Environmentalism' became 'environmental psychology', studying the consequences of environmental manipulations on man (Tuan, 1972); as it happens, geographers are 'hard-put to find a common theory linking . . . geomorphology and psychogeography' (Blant, 1979).

THE 1970S: RELEVANCE, ENVIRONMENTAL CONCERN AND NATURAL DISASTERS

Concern with 'relevance' in the late 1960s encouraged research with bearings on the human condition. Geomorphologists such as Coates (1971) were quick to point out their contributions in applied fields, for instance carrying out studies of accelerated erosion, following sediment losses from coastal and beach zones, river banks, cultivated slopes in drought-stricken areas, and from footpaths in popular parklands. Geomorphologists also contributed to multi-disciplinary schemes in resource management, conservation, and land-use planning for urban, rural and parkland areas. More broadly, the man–land or ecological view wheeled full circle as public interest in environmental questions swung into prominence. In this context geographers were able to claim that their 'particularly valuable asset [was] the continuing viability and vitality of physical geography' (Taaffe, 1974). The irony of geography in the 1970s, however, was that the 'environmentalism' tide of the 1930s had completely run out whereas, in the wake of a gale of technological excesses, it now surged back over so many adjacent strands (Pitty, 1979). Although geographers were among the first to adopt a holistic concept of the environment and its problems and to appreciate the significance of ecology, Macinko (1973) noted that they were 'underrepresented in the recent flood of environmental literature'. From population biology and environmental geology to a range of agricultural and engineering technologies, specialists can obviously provide most valuable insights into ecological, pollution and recycling problems (Taaffe, 1974). The width of interest is evident in environmental geology texts such as *Geology: The Paradox of Earth and Man* (Young, 1975) and *Man and his Geologic Environment* (Cargo and Mallory, 1974). Such views include not only the discovery and acquisition of natural resources, but also the planned utilization of land in rural and urban communities, and the avoidance or control of natural hazards. For example, in flood-plain planning in Chicago, geologists have indicated the needed regulatory and flood-proofing measures from hydrological data (Sheaffer, Ellis and Spieker, 1969). The costs of landslide damage can be high. In Calabria, expenditure for road, railway, aqueduct and housing repairs has been estimated at more than $200 million for 1972–3 alone, with some 100 villages with a total population of 200,000 being abandoned due to landsliding (Carrara and Merenda, 1976). Deeper-seated disturbances are also significant. Apart from the fatalities, it is blunt enough to instance that some 7 per cent of Peru's total population was homeless after the 1970 earthquake. The significance of earthquakes in urban geography is redoubled by their strictly defined localizations. Thus, the most catastrophic earthquake recorded in Europe, at Basle in 1376, occurred at the centre of a dense cluster of recurrent seismic events generated in the area where the Western Alps join the Rhinegraben (Illies and Greiner, 1978). Most cities in Central America are built on flat-lying deposits of pumice or alluvium that fill transverse and longitudinal depressions, located at the intersections of this grid of lowlands. The geographical advantages of these locations are offset by their being loci of moderate-size, shallow earthquakes. Thus, Managua has been repeatedly destroyed by earthquakes, as have the capitals of Guatemala and El Salvador, nine and fourteen times, respectively (Carr and Stoiber, 1977). The social repercussions of the 1972 earthquake in Managua include shanty towns, and the displacement of population and functional changes due to prohibition of reconstruction in the original areas. The geological explanation lies in the continental plate beneath Central America which, due to convergent pressure at its margin, is broken into eight segments some 100–200 km long, by transverse structures. Earthquakes accompany strike-slip displacements on these longitudinal or transverse faults.

Epitomizing the direct relevance of geology to human geography is the evaluation of seismic risk. The implications are serious for the design and siting of nuclear power reactors, dams, underground structures, and further critical facilities for the underground disposal of liquid waste and for hydraulic mining by high-pressure fluid injection. More broadly, the impact of earthquakes on minds may outlast practical responses. For instance, the 1755 Lisbon earthquake, having provided one of many zones of intense shearing between the thoughts of Voltaire and the feelings of Rousseau, is immortalized in *Candide*.

CONCLUSIONS

Although it can be argued that an understanding of social forces requires training, earth and environmental scientists increasingly are urged to examine the social implications of their findings (Lomnitz, 1970). Geographers may again be eager to emphasize the land–man theme, in which environmental forces and social activities interplay, with aspects of both physical and human geography being considered simultaneously. However, scientific emphases of the 1970s question whether this can be designated as a distinctive domain to which geographers have a particular claim. The example of earthquakes illustrates how geologists have a clearer case for emphasizing certain environmental forces. The fact that geology's more emphatic examples originate deep within the earth's crust is a reminder that the land–man view accounts for only a portion of human experiences of natural forces. For those who rise with the sun, sail with the tide, and steer by the stars, the extra-terrestrial component of an Earth-bound existence is far more than the land–man theme, or the etymology of geography, can accommodate.

Statistics and geography

> *The concept of distribution is encountered in all science. In essence it is primarily statistical. It has no exclusive connection with mapping, for distribution can also be expressed in tabular form. (Crowe, 1938)*

INTRODUCTION

The preceding brief sketch on changes of emphasis in geography is sufficient to highlight a major problem in defining the nature of geomorphology. A fine sense of balance is required if geomorphology is to remain accommodated by both geology and geography and yet retain sufficient identity to fulfil a unique function as landform science. This balance is difficult to strike while for geography, at least in America and Britain, there is 'widespread doubt as to what it is all about' (Dickinson, 1969). Quite often, similarities with other subjects are used to support a given author's view of the field, such as economics for Barrows (1923) or history for Sauer (1956). In broad views, the basic feature of geography's complexity is that it spans both natural and social sciences. This is not a unique feature, as it is shared rather obviously with statistics. Therefore, the conceptual juxtaposition of these two distributional sciences might repay brief examination.

As initially employed, the term statistics applied to the comparative description of states, at a time when geography was synonymous with exploration and discovery. Strikingly, the first geographical society in the United States was founded in 1851 as the Geographical and Statistical Society. This body was concerned with both statistical and geographical problems until bifurcation in 1871 (Wright, 1952). Concurrently, in Britain, a continuum between geography and statistics was demonstrated by the scientific and intellectual contributions of Francis Galton. In the 1850s, Galton was an applauded geographer, fascinated by instrumental observation. Although the originator of the 'binary system of exploration', with half the party going back and half going on (Galton, 1857), he challenged the prevailing watchword of geography, 'exploration not education'. He argued that geography 'links the scattered sciences together, and gives to each of them a meaning and a significance of which they are barren when they stand alone'. Later he described ways in which geography needed to be developed and was instrumental in having the subject introduced at Oxford University with Halford Mackinder occupying the first post in 1887. By then, Galton was assuming his role as the pioneer of statistics, commencing with his 'law of deviation from an average', the forerunner of the Normal Distribution Curve on which much of modern statistics is based. In his *Natural Inheritance*, first published in 1889, he outlined 'the law of regression', the precursor of the basic statistical technique that still bears the same name. Together with his friend and former pupil, Karl Pearson, Galton stimulated the remarkable growth in the theory and scope of statistical methods at the beginning of the present century (Fisher, 1963).

GEOGRAPHY AND STATISTICS AS DISTRIBUTIONAL SCIENCES

The fact that statistics, like geography, draws equally from the natural and social sciences is exemplified by the sources from which two pioneers of significance tests in statistics drew their data. In the 1920s R. A. Fisher's attention to agricultural surveys led to his devising statistical methods for studying the simultaneous operation of several variables (Russell, 1966). Equally widely adopted was the 'confidence interval' approach, worked out in the 1930s by J. Neyman from Polish social-survey data. The common feature of statistics and geography is that both are concerned with assessing spatial variability. For one, the concern is with the abstract space occupied by frequency distributions and multivariate associations, the other with the real heterogeneities at the earth's surface. This distinction is not absolute, since geographical areas, beyond a certain scale, become too big to be directly experienced by most people. Thus, 'The region is therefore primarily a construct of thought' (Tuan, 1975). Conversely, statisticians frequently revert to concrete examples of chance mechanisms to illustrate probability theory, such as coin tossing or dice throwing. Both types of distribution are most clearly presented in graphical form. Variability in geography distinguishes between 'near' and 'far' whereas statistics is concerned with 'few' and 'many'. Like geography's multidisciplinary span, with its apparently overambitious width, statisticians also find the same formulae to be equally relevant for widely ranging subject matter. Indeed, both geography and statistics crop up in a complete A to Z of sciences, from anthropology to zoology, with geomorphology adjacent to geology somewhere in the middle. Like the geographer postulating regional boundaries, the statistician is concerned with drawing lines. Thus, the very verge of significance is indicated 'very roughly by the greatest chance deviation observed in 20 successive trials' (Fisher, 1926). In the pioneering of the concept of significance levels in statistics, the 'very roughly' of Fisher's own words cannot be too heavily underlined.

Finally, a more recent development in statistics indicates a further parallel between the two sciences concerned with distributions of data. Most geographers recognize a constraint in regarding their subject as essentially the science concerned with the formulation of laws governing the spatial distribution of certain earth-surface features: this has always been the likely presence in a study area of an unusual, exceptional or even unique case. Although statisticians have also been long familiar with departures from the Normal Distribution Curve and other idealized curves being common in many types of data, it is only recently that the exceptional case has claimed their detailed attention. Significantly, such 'an observation (or subset of observations) which appears to be inconsistent with the remainder of that set of data' is termed an 'outlier' (Barnett and Lewis, 1978). The appropriateness of this geological term for the statistical concept is interesting since, at outcrop, geological outliers nearly always give rise to distinctive, isolated hills. Such landforms are, therefore, very real symbols of the difficulty of searching for generalization in geomorphology.

CONCLUSION

Little may be learnt about the nature of geography from comparisons with subjects which focus on unique subject matter, like the botanists and their plants. Statistics, however, demonstrates that a distributional science's material is clearly drawn from a variety of other studies. Disciplinary viability does not necessarily depend on distinctive subject matter. Despite their repeated exercises of flipping and throwing, statisticians lay no particular claim to coins, dice, or to extracting balls from bags. In the enormous range and variety of interests in which statistics crop up, it is evident that no one theme is either identifiable or indeed necessary, and that a loosely knit range of specialisms is needed for differences in data, purpose, methods and styles of enquiry of particular specialist sciences. Although correlation methods are perhaps the most widely used, causality is never assumed. Indeed, statisticians tend to leave the interpretation of associations, judged 'significant' by their methods, to the expert in the given field.

The comparison of statistics and geography suggest that, essentially for the present purposes of elucidating the nature of geomorphology, the following general working definition of geography might be employed:

> Geography describes macroscopic or abstract flows and the distribution of visible forms at, or close to, the earth's surface, and the character of places where such flows and forms intermingle. Analysis may be emphasized where patterns or covariations are found to be present or distinctive individual cases identified.

In this working definition, it is implicit that flows are spatial phenomena. No reference is made either to whether form or flow might be considered part of a natural or a social science, or whether interrelationships might be considered only between or within either setting. Forms may be natural and as fleeting as clouds or artificial, geometrical and as enduring as Stonehenge. Flows may be abstract, as Hägerstrand demonstrated in his pioneer work on the diffusion of ideas in the 1950s, as quick and as idiographic as Concorde, or as slow, inexorable and universal as continental drift. The working definition avoids the assumption that geography has some special claim to study human inter-actions with the environment without limiting its possible adoption as a central theme. In particular, it avoids the pitfall of causality being a necessary condition for holism, or that disciplinary coherence necessitates 'welding' or 'fusion' by tenuous or contentious assumptions. Expression of the interaction between flow and form may range from scales of the Aswan Dam and Himalayas to those of striations and graffiti. For the future of a more narrowly defined geography, the writing is perhaps already on the wall.

The fourth dimension

Formerly, fixistic geology was concerned with up, down and time, but now we must add sideways. (Dietz, 1977)

With resignation, geographers often admit that whether geography has emerged as a natural science or a social science 'has been a problem of administration rather than of scholarship' (Finch, 1939). It would be helpful to have an operational approach that was not channelled along corridors of power, avoided too much reference to the specialist sciences into which segments of knowledge are conveniently grouped, and avoided any narrowness in following one particular school of thought. Some progress in this direction can be made simply but effectively by considering the priority in which a given specialist view ranks the four dimensions of length, breadth, height or depth, and time which make up the spatiotemporal structure of the real world. For example, it is over spans of time that the historical science of geology is primarily spread. As time is physically manifest in the stratigraphical column, the vertical is the first of the three spatial axes in most geological sciences. With strata commonly inclined, it is width of outcrop which is then of particular interest, with cross-sections essential supplements to geological maps. On larger scales, widths express the dynamism of the plate-tectonic model, explicit in the concept of 'sea-floor spreading'. Lengths are commonly less critical in geology, although vital in mineralization of lodes and in the exorable movement in strike-slip faulting.

Geography, however much the present day needs to be seen in the context of the receding past and approaching future, is concerned by definition with the spatial dimensions of the modern world. With journeys to work, worship, well or washplace, and other frictions and tyrannies of distance, length is the geographer's primary dimension. Napoleon, an arch exponent of utilizing knowledge of terrain, might have supported this assertion more than most, whether from Moscow or Elba. With portrayal in maps being an aid and expression of geographers' work, width automatically follows as a secondary dimension. Since humanoids first climbed down from trees and got their feet firmly on the ground, the vertical has become the least significant spatial dimension and, with the possible exception of aspects of transport geography (Appleton, 1963), is often neglected in many human geography studies. Indeed, in maps showing patterns and distributions in the past, the vertical becomes the fourth dimension.

By definition, the study of landforms is primarily concerned with the present ground-surface shape, in its three dimensions. The vertical is the primary axis because, in addition to being a spatial dimension, it is also an approximate index of energy available for natural transportation processes. Length is the secondary spatial dimension since most natural transportation routes are gravity-driven, terrain-channelled trends or 'clines'. Thus many landforms are expressed as cross-sectional profiles, as in stream, slope or beach profiles. In the chronological sequence of such profiles, width is a fourth dimension.

This exercise in ranking the priority given to dimensions does not, of course, imply that a given 'fourth' dimension is insignificant. In fact, it highlights a source of problems in many studies where the fourth dimension is inadvisedly ignored, such as living conditions in tower blocks or the width of Horton's belt of no-erosion in watershed divides. Its validity can be briefly checked by considering its application to studies related to geomorphology and geography. In pedology, for instance, a ranking of dimensions appears equally clear-cut. Soil properties change dramatically down the soil profile and so the vertical, readily investigated in the field by auger, is the primary dimension. On broadest scales, latitudinal widths of major soil types caught the eye in Russia in the 1880s and, on local scales, the width of soil type in downslope changes is the basis of mapping. There are not many references to how 'long' a soil is, although the 1 m^2 *pedon* is a three-dimensional extension of the soil profile in detailed soil taxonomy. Soil-profile development reflects a balance with the environment, and nearly steady-state conditions are assumed, with additions from weathering and decomposition of organic matter assumed to equal approximately soil losses. In consequence, pedologists may mention time rarely. In hydrology, rain falls, streams run off, and moisture is returned to the atmosphere by evapotranspiration. Again, the vertical is the primary dimension, as seen in the hydrological cycle. The lengths of flow, however, are influenced by the time when rain falls, or snow melts, or when evapotranspiration is least. Whether for flood or drought, time is critical, and length might be seen as a subordinate dimension. Given permanent, normal flow, streams' length expresses the control of gravity on drainage. Width only becomes a critical dimension when bankfull stage is exceeded. Finally, in the abstract world of statistics, the first two dimensions of a population's distribution are clearly ranked, with the relative frequency of a measure of central tendency, expressed vertically, being the first consideration. Estimations of the distributions width then follow, as the standard deviation or interquartile range. Although multivariate statistics operate in an abstract multidimensional space, by far the most common third dimension explicitly mentioned in statistics is time, since so many critical levels of significance in economics, agricultural, or meteorological statistics depend on time. If lengths are mentioned in statistics, they usually refer to spans of time.

If this particular suggested ranking of dimensions is disputed, the disagreement should be taken as a working example of the source of many apparently intractable debates where the basic issue can often be identified as rival views simply emphasizing different dimensions. Intolerance in such debates is not so much an unwillingness to see the others' point of view as a rejection of their primary dimension. Also, an exchange of third- and fourth-order dimensions may be advantageous in particular regional settings. Thus, the longer history of persisting settlement makes time a more prominent dimension in geography in the Old World than in the New.

When the suggested ranking of dimensions is tabulated (Table 2), several possible comments are invited. First, it indicates how a specialist in one field may make a contribution in an adjacent area. For instance, a pedologist focusing on a relict soil, as in examples summarized by Birkeland (1974), converges on geology by putting time as a primary dimension. Conversely, if a sedimentary petrologist wishes to examine present-day analogs to an ancient sedimentary environment, even if the ultimate purpose is the elucidation of geological time, the dimensions of the shorter-term priority suggest that the

Table 2 Postulated priorities in which specialist studies rank the four dimensions of reality

Earth sciences				Distributional sciences	
Geology	Hydrology	Pedology	Geomor-phology	Geography	Statistics
time	vertical	vertical	vertical	length	vertical
vertical	time	width	length	width	width
width	length	length	width	vertical	time
length	width	time	time	time	length

study will equally be a contribution to geomorphology or pedology. The emphasis on time, implicit in hydrology, may explain why this study is often so efficiently accommodated within geological institutions, as illustrated visibly by the length of shelf space occupied by the Water-Supply Papers of the US Geological Survey. The validity of the ranking is perhaps supported by the similarity of pedology and geomorphology, in view of their strict juxtaposition in nature. Geography stands out as the science that is primarily concerned with length. Also, the emphasis on the vertical in geography is less than in any of the other sciences considered.

The ranking of dimensions suggests that geomorphology, although close to soils, is rather different from both geology and geography. When the geologist uses the word 'length', the reference most commonly will be to spans of time. In contrast, length is one of the two main dimensions of geomorphology and also the main axis of geography. However, vertical height, the main dimension of landforms, is inevitably the subordinate spatial dimension of geography and is, in practice, often discounted. Perhaps geomorphology's distinctive but intermediary position between geography and geology might be summarized by the following definition: geomorphology is the geographical science that examines landform, the physical property of rocks due to their exposure in the earth's subaerial environment.

Conclusion

Landform, as the ground-surface expression of rocks, is particularly important in reconnaissance and basic training for geological mapping. Equally, the more obvious geographical changes in landforms are often attributable to contrasting geology. Such changes direct the course of many closely related natural phenomena, such as soil changes and drainage, and influence the cost of many human endeavours. However, it seems that even where geologists and geomorphologists study the same form, process, stratigraphical record, or contemporary environmental problem, differences are usually discernible in purpose, perceptions and weighting. The order in which the four dimensions of the real world are ranked by definition or implication is helpful in discriminating between the emphases of either geography or geology. Thus, a geomorphological enquiry that emphasizes time contributes to the last chapter of the geological record; one in which lengths are prominent is particularly geographical.

The necessary disciplinary exercise of differentiating the geological from the geographical need not become the overscrutiny of a false antithesis. The interdisciplinary emphasis of the 1970s on research relevant to the plights and comforts of the human condition has reminded and revealed that many strictly geological phenomena have a direct bearing on human geography. Because most of these geological phenomena are equally significant in geomorphological studies, a geography incorporating geomorphology keeps open a channel through which such geological information can be incorporated into rounded geographical studies. Therefore, both in terms of areal variation and the man–environment theme, the paradoxical conclusion emerges that it is the geomorphology emphasizing geological foundations which has most direct bearing on the study, training and practice of geography and geographical methods. This role could be strengthened by using more geomorphological examples from areas which have considerable human interest as well, rather

Figure 3 Geomorphological history of an area near a major urban centre—the ancient course of the Tiber River near Rome
A   Pliocene paleogeography, with shoreline much to the east of its present position, and only present-day, left-bank tributaries in existence.
B   Early Pleistocene, after uplift and the formation of intermontane lacustrine basins and the emergence of the lower Tiber region, with left-bank tributaries linked to the Cromerian 'Paleotiber' trunk stream.
C   During the Pleistocene, intermontane lakes gradually dried up, and part of the Tiber drainage was diverted into the Arno by regional tilting.
D   Emplacement of tuff following the first eruption of the Sacrofano volcano, and the formation of lake beds in the obstructed 'Paleotiber' valley in post-Cromerian/pre-430,000 yr BP times.
E   The dammed lake backed up the northern tributary as far as Torrita Tiberina, where the divide was breached by spillover from the lake. With gradual drainage of the lake, the present course of the Tiber was completed.
Source: Adapted from Alvarez, 1973.

than too many citations of inaccessible niches remote from civilization. Thus, Figure 3, illustrating drainage diversions near Rome, emphasizes both the geological basis of geomorphology and equally the occurrence of a classic geomorphological phenomenon close to a long-settled city.

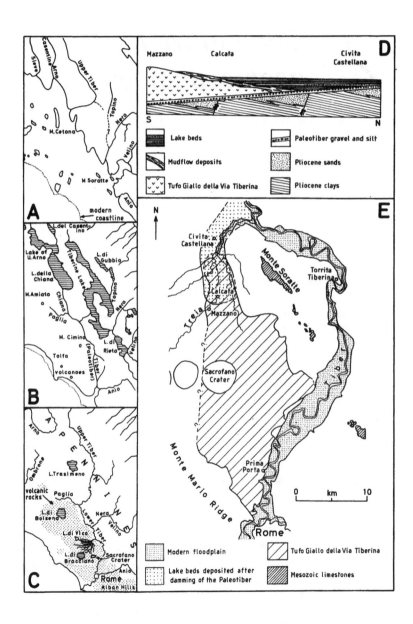

2 The nature of geomorphology

Description and interpretation

Any serious student of geomorphology will quickly realize that what is actually known with certainty about landforms and their origin is surprisingly small. (Small, 1978)

WORDS

Any science starts with observation of natural phenomena. The first system-atically collected observations of landforms during the last quarter of the nineteenth century were characterized by lengthy verbal descriptions. Gradually, sufficient symmetry and distinctiveness were recognized in certain landforms for their classification as one of a certain type. Brief terms, such as river meander or bay-head bar, conveyed large amounts of descriptive infor-mation. However, the designation for a particular feature is not always agreed. Furthermore, the number of shades of meaning lengthen as the literature grows, as noted for the term 'pediment' and the French equivalent 'glacis' (Whitaker, 1979; Tricart, Raynal and Besançon, 1972). Indeed, the exact translation of the innumerable local terms can be a persistent problem until a local expert describes in detail the exact connotation, as in the case of the unique karst landform, the 'polje' (Gams, 1978).

Genetic description may afford further economy in words, with such terms as 'storm beach' implying interpretations of origin as well as describing the appearance of forms. Indeed, geomorphology is unusual in the degree to which descriptive explanatory or genetic terms have been employed. They are particularly common for landforms confidently attributed to glacial action, such as corrie, drumlin, or moraine. The corrie is also an example of a local descriptive term, the Scottish equivalent of the Welsh 'cwm' or the French 'cirque', which has taken on a genetic meaning. The trend since 1945 has been away from genetic description, to separate clearly description from interpretation. Thus, Dury (1972) recommends that the 'pediment' be defined in terms which are independent of climatic and process connotations as 'A degradational slope, cut across rock in place, abutting on a constant slope at its upper end, and decreasing in gradient in an orderly fashion in the downslope direction'. Previously, the most controversial have been genetic terms which, in addition to describing and explaining, also imply relative age of forms. This triple economy characterized the terms of W. M. Davis's Cycle of Erosion, such as rejuvenation, peneplain, and river capture.

The appeal of the brief, vivid, and perhaps poetic metaphor has led both popular and technical writers to animate the landscape (LeGrand, 1960). Analogies with the human body are a veritable identikit. Descriptive terms include 'headlands', 'lobes', 'mouths', 'gorges', 'volcanic necks' and 'elbows', down to the 'cold foot' with 'glacier toes and soles'; not least, there is 'the skin of the earth'. Although the metaphor and the analogy were commonly used devices in explanations by Davis and others who have relied largely on verbal descriptions of phenomena, the mental picture which metaphors invoke may be unrelated to the dynamics of a process. Further enquiry into the mechanics of apparently self-explanatory headward erosion reveals little beyond the assurance that it breaks, bites back, or even gnaws into a divide. Sometimes, ill-chosen analogies are unmistakably applicable in one or two senses only, and the many differences between the geomorphological phenomenon and the analogy drawn may become distracting. A table land laid with hogsback ridges and basket-of-eggs drumlins would make a fine geomorphologists' breakfast, before turning to the washboard moraines and sinkholes and then relaxing in armchair hollows.

Analogies are commonly a persuasive basis for mathematical models, and careful screening of meaning is essential before getting enmeshed in formulae. For instance, early estimates of the viscosity of lava 'streams' were based on the analogy between these flows and that of river water, and the applicability of hydraulic equations to lava was mistakenly assumed. Also, the analogy with organisms that flourished in post-Darwin years (Stoddart, 1967) but was becoming an endangered species of specious argument, has been revived as 'allometric growth' (Woldenberg, 1966): as applied to an animal, the law of allometric growth states that the relative rate of growth of an organ is a constant fraction of the relative rate of growth of the whole organism. Size-related differences in landforms, however, are inanimate and better described simply as power-function relations (Mosley and Parker, 1972).

Since many explanations are triggered by visions of a link between two apparently unrelated phenomena, the analogy can be vital in research. The analogy and the metaphor are integral parts of human communication, the wheels of the vehicle of thought. Therefore, 'Since metaphorical usage is inescapable, it must be more explicitly articulated. Since it has explanatory value, it must be consciously encouraged' (Livingstone and Harrison, 1981). Such care is vital to geomorphology, perhaps one of the most verbal of sciences. Care is needed, not least with everyday words, particularly where criticism is levelled, or where anthropogenic roles are considered.

SKETCHES AND PHOTOGRAPHS

Before photography became popular and inexpensive, landscape drawings and sketches were the conventional visual description for landforms, the now-treasured ornaments of nineteenth-century earth science. Davis, however, drew not merely depictions of relief, but his sketches incorporated interpretations too (King and Schumm, 1980). Other pioneers also found that, by subtle accentuations, explanatory model sketches were persuasive supports for their hypotheses. Once the counterparts to the contemporary whizz-kids at their computers were the sketching wizards at the blackboard, constructing three-dimensional sketches, using both hands at once. Today, diagrammatic drawings of idealized composite views are established features in basic textbooks. The field sketch can remain an elementary exercise in which students can record their appreciation of significant geomorphological features because the increased attention to process and sediments can be incorporated into the traditional morphological approach (Fig. 4). It is significant that the art of drawing fossils is valued in palaeontology.

In emphasizing pictorial presentation, geographers move with the mainstream of contemporary culture, which is heavily visual (Tuan, 1979). The photograph is obviously an attractive aid, depicting a near-horizontal, two-

---

Figure 4 Sketches illustrating some significant factors in contemporary studies of fluvial geomorphology, as observed in central Texas
A   The influence of lithology on sedimentological changes (i), and contrasts between channels in limestone (ii) and granite (iii).
B   Effect of increasing flow on channel-perimeter roughness.
  (i) Non-monotonic relationship between depth of flow and roughness. Letters a, b and c represent this relationship at the flows illustrated in the sketches (ii), (iii) and (iv) respectively.
 (ii) Rough channel at low flow amid boulders, decreasing as stage rises to bankfull.
(iii) Overbank stage results in increased roughness.
(iv) About flood peak, roughness is less, due to evacuation of boulders and the uprooting of vegetation.
C   Gravel berm and other deposits on the inner bank of a meander, the results of a May 1972 flood in Bielders Creek, near New Braunfels, when a peak discharge flowing at 2.5–3.3 m/sec. followed a 4-hour rainfall of 400 mm.
Source: Adapted from Baker, 1977.

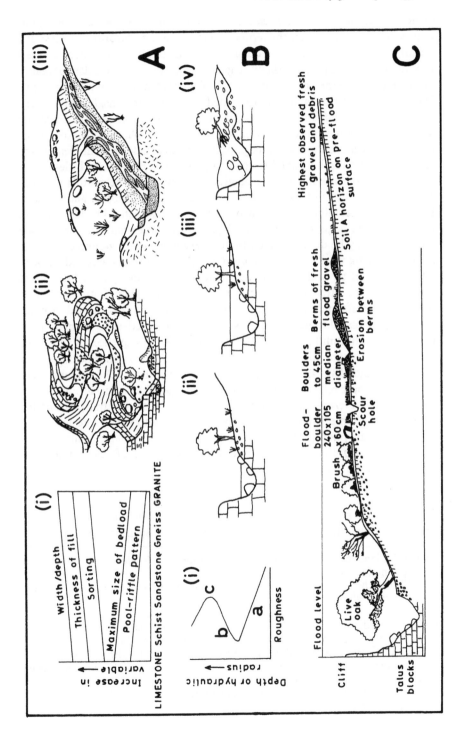

dimensional view of landforms, a conveniently filed, pocket-sized replica of reality. However, because the reproduction is so exact, the distinctiveness of a locality is portrayed in all its detail. Such idiographic knowledge, focused literally on the unique, is difficult to generalize. Also, visual displays of reality do not encourage the habits and patterns of abstract thinking in which generalizations are construed and comprehended, and where the imagination ranges.

The practical advantages of aerial photographs are important since the stereoscopic image of vertically taken, overlapping photos is the object of landform research. Prior knowledge of the terrain of field areas and features of special interest can be gained from air photographs, if large and little-known areas are to be investigated. Geological influences can be postulated, particularly structural forms. Fault traces and scarps, dipslopes, and volcanic forms are discernible, as are major joints and associated breaks of slope, rectilinear drainage patterns and depressions. Potentially unstable slope conditions may be detected, a particular advantage where ground conditions for engineering works are being considered (Norman, Leibowitz and Fookes, 1975). Also, if an area lacks an adequate geological map, some lithological boundaries are often evident on air photos. Floodplain features such as meander cutoffs and levees are readily identifiable, as is patterned ground, including areas of relict permafrost. Lighter-coloured subsoils may show through in subhumid areas affected by erosion. Time sequences of air photos are valuable where surface forms change rapidly, as on snowfields, glacier surfaces, braided stream channels, beaches and coastlines. The limitations are that stereoscopic relief is two to three times steeper than actual ground slopes and becomes increasingly distorted outwards from the central point of the photograph. Some surface forms may remain undetected in forested areas. Ultimately, air-photograph interpretation yields most reliable results if summarized as a preliminary map and then the nature of selected features examined in the field.

MAPS

The contour map is a much-used aid in many types of work. For geomorphologists, the contour map is the basic, indispensable description of

---

Figure 5 Plane-table contour maps as a fundamental, detailed, three-dimensional description of landform
A   Hillside channels in the Sanfjället area of Härjedalen, Sweden, which Mannerfelt (1945) attributed to sub-ice drainage.
B   A barchan dune near the Salton Sea, California. Despite the close contour interval, it was reported difficult to estimate slope angle adequately from this detailed map.
C   Typical example of prehistoric earthflows on the Nyika Plateau, a high-altitude grassland in north-central Malawi. The 0.5 mm^3 volume of the slipped material, with hummocky relief subdued by age, maintains a marsh upstream and rapids downstream in the North Rumpi stream at its toe.
Source: Adapted from A Pitty, Foster and Foster, 1976 (unpublished); B Howard, 1977; C Shroder, 1976.

their subject matter, portraying the vertical dimension precisely, and outlining vividly the length and breadth of landforms. Block diagrams may occasionally be drawn from a contour map, with or without perspectives but preferably with computer. Such diagrams convey more precise information than photographs or sketched block diagrams.

Landforms of particular interest are often too small to feature on commonly available map scales and contour intervals. Thus, a laborious but rewarding

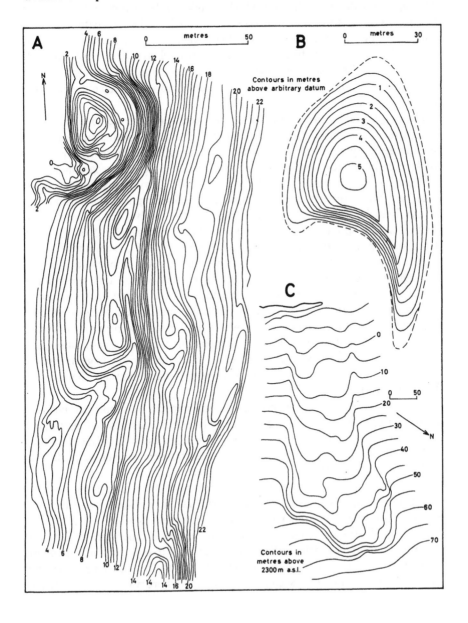

exercise is to survey small landforms at appropriately narrower contour intervals. For example, Figure 5A shows the typical details of the type area in Härjedalen in central Sweden from which Mannerfelt (1945) developed his ideas on the downwasting of ice-sheets. His interpretations have been tested successfully in other areas, particularly in Britain, although there are difficulties in following Mannerfelt's lengthy exposition in his native language. The detailed contours from this area of minimal contemporary surface drainage and deep regolith on which boulder-glide furrows are found, are helpful in suggesting that one type of 'channel' is the crescentic scar of a landslip. For instance, the 'up-and-down' channel at the north-west end of Figure 5A shows local obstructions on the 'channel' floor, not consistent with water action. Less controversially, the barchan dune and the earthflow in Figure 5 exemplify the quality of the contour map as a highly efficient international language.

On larger scales, and particularly where contour maps are not available or too coarsely contoured to represent sudden changes in ground-surface slope, standardized symbols can be plotted. Such simplified representations of the surface forms were initially popular with several American geographers who enlarged on the pictorial symbols used by W. M. Davis in his block-diagram sketches. Subsequently, the interest in devising an acceptable standardized set of symbols for morphological mapping was the catalyst around which the British Geomorphological Research Group first congregated in 1960. Distinct from such purely descriptive mapping of form (Fig. 6), exemplified by the work of Savigear (1965), other European schema usually embody decisions on the age and dominant process of landforms. However, for investigating specific landforms in a particular area which is not mapped in its entirety, parsimonious use of symbols is a clear advantage and a local scheme of symbols employed (Fig. 6B). The degree to which local distinctiveness may override advantages of a general scheme is demonstrated by soil classification and mapping preferences. Similarly, therefore, a satisfactory key to geomorphological maps may be developed for specific regional conditions (Barsch and Liedtke, 1980).

One immediate way in which the contour map can be supplemented is by plotting a profile. River valley profiles may be either transverse or longitudinal in cross-section and, like the slope profile, greatly assist the visualization of relief. Valley profiles are more helpful in describing and explaining landforms if the geology along the plane of the profile section is included.

Apart from the contours themselves, shapes of drainage basins can be outlined on maps by lines following watersheds. More extensive use has been made of the blue-line drainage pattern itself, largely as a basis for ranking a hierarchy of stream orders. Intending to improve runoff prediction, the American engineer Horton (1945) assigned a rank to each stream, depending on its relative position in a drainage network. Thus, first-order streams occur above the first confluence, second-order streams where two first-order streams

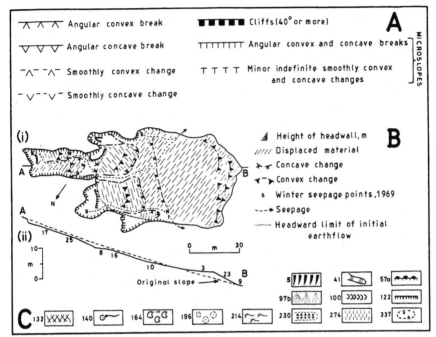

Figure 6 A selection of landform mapping symbols

A   Some slope-form mapping symbols.

B   Example of a morphological map, featuring an earthflow in the Chittering Valley on the Darling Scarp, north-east of Perth, Western Australia.

  (i) Map showing form of displaced material, much of which remains within the flow track. Seepage characteristics are recorded since throughflow above an impermeable subsoil is a critical factor in materials in which shear strength decreases rapidly with increased moisture content.

 (ii) Important supplementary information on a cross-section (A–B on the map), demonstrating the threshold slope of 15°, essential for earthflow in this area.

C   Diversity of symbols in the unified key for detailed geomorphological mapping of the world. The examples illustrated are:

5 thrust scarp
41 lava flow
57a edge of mesa (sandstone)
97b alluvial fan (sand)
100 delta levées
122 ice-contact slopes
132 limestone pavement
140 swallow hole
164 roche moutonnées
196 nivation hollow
214 turf-banked terraces
230 rock glacier
274 loess-cover
337 mining subsidence basins

Sources: Adapted from A Savigear, 1965; B Pilgrim and Conacher, 1974; C Demek, 1972.

combine and third-order streams where two second-order streams join. Horton's procedure of then tracing back the highest order in the basin to the source has been replaced by the Strahler system of ordering (Fig. 7A). However, stream ordering remains a rather crude classification of streams and may not be sufficiently sensitive to express drainage-basin geomorphology adequately. One reason is that the addition of lower-order streams need not increase the order of the main channel. Drainage networks may be described more comprehensively in terms of lengths and connecting patterns of 'links'. Terms include 'sources', at the tips of unbranched tributaries. A 'link' extends from either a source or a junction at its upstream end to either a junction or an outlet at its downstream end (Shreve, 1966). By definition, no tributary joins a channel within a link. Exterior links stem from sources; interior links stem from junctions. Interior links may be subdivided, depending on whether the tributaries at their ends join from the same side (cis-links) or from opposite sides (trans-links) (James and Krumbein, 1969). If geomorphological sub-divisions at a regional scale are to be of geographical value, they must be based on spatially meaningful attributes of landform rather than on spatially random attributes such as network topology (Gardiner, 1978). Since both stream ordering and link analysis dispense with geographical orientation, stream junction angles between confluent stream links describe the spatial structure of the drainage network (Fig. 7C). Thus, Jarvis (1976) has established precisely the degree of clustering of orientation of streams confined within glaciated troughs in the Southern Uplands of Scotland.

Although an indispensable aid in describing landform, a map itself can support little interpretation. Indeed, there is a temptation to read too much from a small-scale map at second hand and even with a digitizer. As Davis cautioned, map study 'seems to lead different investigators to different results'. Map availability may also lead on research rather than follow it. Certainly the converse has been suggested for Australia where limited availability of detailed contour maps may have restricted the spread of erosion-surface mapping (Spate and Jennings, 1972).

IRREGULARITY AND COMPLEXITY OF GROUND-SURFACE SHAPES

Maps with widely spaced contours may smooth over the fact that the objects of geomorphological study range from the regularity and symmetry of the barchan dune to amorphous rock outcrops on ancient shields and jagged, glaciated highlands. In the term 'stream order', 'order' is optimistic. Even the barchan dune is slightly disordered, with one horn usually longer than the other, and, in erosional forms, the symmetry of the corrie or the regularity of the pediment is an exception rather than the rule. Admittedly, the more regular and symmetrical a landform, perhaps the more there is for the geomorphologist to explain. Since much of the inclined landsurface is intrinsically complex, having little or no regularity of shape, such generally amorphous

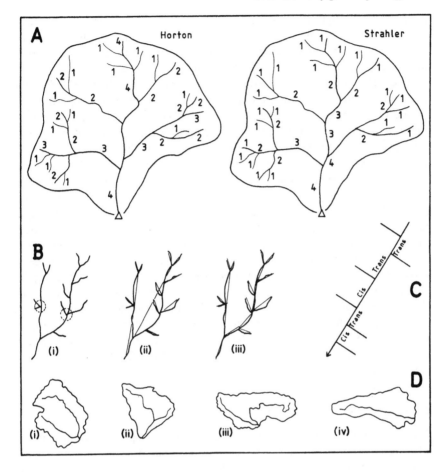

Figure 7 Some definitions used in describing stream-channel networks and drainage-basin shapes
A   Stream-ordering systems for describing drainage composition.
B   Stream orientation.
 (i) Problems of closely spaced junctions in the resolution of stream reaches by Lubowe (1964).
 (ii) Over-generalization in resolving a stream reach containing an unrecorded number of tributary junctions in a single vector.
(iii) The stream-link vector, as the straight line joining the two bounding junctions of the link for interior links, and joining the source to the first junction for exterior links.
C The James-Krumbein (1969) classification of link topology.
D Various drainage basin shapes; these examples are from the Appalachian plateau, with corresponding indices listed in Table 3.
Sources: Adapted from A Bowden and Wallis, 1964; B Jarvis, 1976; C Jarvis, 1977; D Morisawa, 1958.

landsurface configurations should be clearly distinguished from *landforms*, in which the recognition of some degree of regularity is implied. This distinction is seen most clearly in the complexity of coastal development, with its diversity of features from jutting headland cliff to smoothly arcing bay.

The difficulties posed by the inherent irregularity of much of the landsurface is well illustrated by ambiguities in descriptions of the meander loop, since this is one of the most regularized of forms. Indeed, the radius of curvature is nearly constant for a full third of the length of the meander loop on large lowland rivers. However, a meander loop is not usually paired with another loop of the same size and form in natural streams (Brice, 1974). River-channel 'sinuosity' has been defined as the ratio of channel length to valley length, although valley length is not readily defined exactly. In fact, the overall trace of a meandering stream may be so tortuous that measuring a meander wavelength is impracticable. A major additional but less well-understood characteristic is variation in the down-valley configuration of channel floors, as deeper pools alternate with shallower riffles. There are further irregularities in channels in cross-section, with the main idealized parabolic, trapezoidal, or rectangular sections overestimating the ease with which even stream width might be defined. Although the water-surface width may increase progressively with increasing discharge, effective width of active flow may change abruptly (Richards, 1976). Even the most clear-cut width, at the bankfull discharge, can change with scour-and-fill. Definitions may need modification if banks are of unequal height.

MEASUREMENT OF CO-ORDINATE AXES AND RATIOS

Measurement is usually a preferred mode of description, but such is the ill-ordered configuration of the landsurface that it is not always readily amenable to numerical description. It is difficult to characterize an irregular, natural ground-surface shape by simple, single parameters, and no single dimension can describe the ground surface entirely if used alone. Efforts to summarize landform in numbers are usually based on ratios between various dimensions of the form under consideration.

The problems of defining measures of landform in plan are well illustrated by representations of drainage basin outlines (Fig. 7D). Horton suggested a ratio of length of drainage basin, L, to its width, W. Length is the longest dimension from the outlet to the opposite side, with width measured normal to the length. S. A. Schumm suggested the index $E = d/L_m$, where E is an elongation ratio between d, the diameter of a circle with the same area as the drainage basin, and L_m, the maximum length of the basin parallel to the main river. A similar index is $C = A_b/A_c$, where C is J. P. Miller's Circularity Ratio between basin area, A_b, and the area of a circle, A_c, having the same perimeter. Table 3 lists calculations of these indices for the basin outlines shown in Figure 7D. However, areas computed from a hypothetical circumference are not a sufficiently accurate way of measuring shapes of drainage basins. Several differently shaped basins might have similar indices due to irregularities in their outline, and the Circularity Ratio describes increased irregularity of the basin outline rather than basin shape. Equally, the measurement of basin length is problematic when the long axis of the basin trends away from the line between basin outlet and the opposite side of the basin. Perhaps the most useful readily calculated ratio for drainage basin description is drainage density, since this reflects energy inputs and because runoff and sediment yields are influenced by length of water course in a given area.

Table 3 Indices of drainage basin shapes (from Morisawa, 1958)

Drainage basin	(i)	(ii)	(iii)	(iv)
E	0.82	0.86	—	0.59
C	0.58	0.64	0.47	0.45
L/W	1.32	0.97	0.50	2.17
area (km^2)	66.60	48.40	222.70	27.50

The range of examples of other forms in Figure 8 starts with the definition of a weathering recess, partly because the simplicity of the index belies the infrequency with which such measurements have been recorded. With the greater irregularity of erosional forms, like the nivation hollow, several measurements might be made to avoid ambiguities. Figure 8 also illustrates the spacing and consistency of orientation of some landforms. Although the geometrical regularity of drumlins makes these forms an obvious example, measurement of spacing of more irregular forms is feasible.

Cross-sections and profiles can be illustrated by the case of the beach profile in which the *x*-axis extends seaward perpendicularly to the shoreline at mean sealevel, and the *y*-axis extends vertically upward. Although amongst the smoothest of landform profiles, other parameters are needed to define a beach profile uniquely (Sonu and van Beek, 1971). Contrasts in profile shapes depend on whether the surface configuration is curved concavely upward, is straight, or curves convexly upward and, if a berm is present, whether the berm occupies the lower, intermediate, or upper portions of the beach. For hillslope profiles the ratio between *x* and *y* axes is simply the equivalent to mean slope angle, with the configuration specified by a downslope sequence of several slope angles. For stream longitudinal profiles certain segments are often sufficiently smoothly curved to have invited description by mathematical formulae, particularly in homogeneous lithologies like chalk (Culling, 1956). More commonly, changes in rock outcrop produce irregularities that are discontinuities comparable in relative scale to berms on beaches. In consequence, an adequate expression of overall concavity can be obtained from two lines drawn on a stream longitudinal-profile plot, one joining the headwater point to the estuary or tributary junction downstream. The second line is

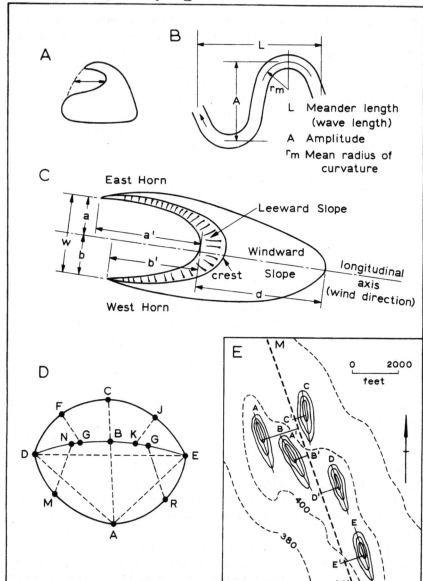

Figure 8 Examples of some measured dimensions in more symmetrical cases
A   Depth of cavernous weathering.
B   Meanders.
C   Barchan dunes.
D   Nivation hollow. DCE is the top of the backwall, DBE the base of the backwall, and DAE the front of the hollow floor.
E   Drumlin spacing and orientation.

Sources: Adapted from A Calkin and Cailleux, 1962; B L. B. Leopold, M. G. Wolman and J. P. Miller, *Fluvial Processes in Geomorphology*, W. H. Freeman & Co. © 1964; C Finkel, 1959; D Cook and Raiche, 1962; E Reed, Galvin and Miller, 1962.

parallel to the first, but as far from it as possible while still being in contact with the profile configuration. An index of concavity is that of the perpendicular interval between these two lines, divided by the height of the plot, and is sufficient to highlight striking regional comparisons and lithological controls (Wheeler, 1979). Such dimensionless indices are generally useful because they do not depend on the scale of the feature and are widely used in sciences closely related to geomorphology. For instance, the comparison of floods in hydrology is facilitated by the unit hydrograph. A given storm runoff is plotted and then the ordinate of the storm hydrograph plot is adjusted to generate a hypothetical . hydrograph corresponding to 1 cm, or 1 inch, of rainfall.

CONCLUSIONS

Morphometric data provide no simple solutions to pediment problems.
(Cooke, 1970)

In the last twenty years geomorphologists have moved away from genetic descriptions, attempting to separate more clearly description from interpretation. However, it remains that descriptions mean little in isolation from the context of explanation of which they are a part, and that description and interpretation are not as sharply distinguishable as convenience might prefer. The probability of some ambiguity in defining the units of measurement must also be anticipated. Inevitably, the line of mean sealevel on the beach, the position on the stream long profile where the estuary starts, or where in the valley floor the slope profile stops, may not be exactly identifiable as specific points. In stream ordering, workers may report difficulties in following a definite channel in the field or in finding its corresponding representation on the map. Ongley (1968) is only partially correct when he states that 'the intuitively simple concept of basin axis . . . is one of the most difficult geomorphological properties to identify precisely and objectively'. This level of difficulty is simply typical. Therefore, the efficiency, limitations and exact meaning of every measuring definition and index in geomorphology must be carefully evaluated at the outset.

Explanation in landform studies is confronted by formidable difficulties which in part reflect the problems of achieving a precise basic description of even simple forms. Another major problem in searching for explanations of landform is the intrinsic incompleteness of the evidence which may be insufficient for drawing firm conclusions. A further problem in the study of interadjustments between landform and natural forces changing continually through time is that isolated causes are usually inseparable from distinct effects. All that can be recognized is a distinction between the antecedent conditions which affect subsequent conditions. In addition to the incompleteness of the evidence, the sheer variety and intricacy of interrelationships between biological, chemical, physical factors, and antecedent landforms means that there are few laws that are invariably applicable, as in physics, on which to base interpretations of much of the present-day landsurface configurations. The insufficiency of physical conditions to specify uniquely the result of the interaction between dependent variables has been termed 'indeterminacy' (Leopold, Wolman and Miller, 1964; 274). There is no prospect of *proof* in the absolute sense as in mathematics. There are no *chemical formulae* such as those which specify igneous rocks or characterize soil horizons. There is no *genetic code* like that which links plants and animals in time and space. Geomorphological explanations, therefore, tend inevitably to be qualified by such phrases as 'tends to' or 'possibly'.

If one further word might be added to geomorphology's overburdened vocabulary, it is the Japanese word *mu*. This word indicates that the context of the problem is such that to answer either 'yes' or 'no' would be equally inappropriate. Thus, in presenting a theme on description and interpretation, the overlapping of these two conveniently designated end points to a spectrum of reactions of thought and action is stressed. Equally, to avoid polarization into a false dichotomy, a third angle is stressed and exemplified, that of imagery and analogy which permeates both but is identifiably neither. Just as there are more than two dimensions to landforms, so there is often a third if not a fourth way of dissecting the several dichotomies on which the nature of geomorphology appears to be structured. Thus, James (1967; 19) has stressed that

> among the persistent oversimplifications, we should provide a special place for the persistence of dichotomies. By their simplicity and apparent clarity they have the effect of ending critical thought. They provide the tired mind with clear-cut categories with which to become associated; but perhaps nothing is so pervasive and so ambiguous in promoting the persistence of error.

He also advised that 'It might seem advisable to avoid the intellectual arrogance suggested by too much emphasis on explanation' (James, 1967; 4). However, some progress has already been made in drawing together the ends of the major dichotomy which geomorphology partially straddles between the natural and social sciences. Even within physical geography, geomorphology is less governed by the laws of physics than other components such as hydrology and climatology. In social sciences, 'subject matter is remarkable neither for the relative simplicity of pre-atomic physics nor for the organic integration of physiology' (Homans, 1967). There is also a strong element of historicity, with past circumstances combining with the present to explain behaviour. Thus, if much of the detail of geomorphology appears 'indeterminate' when approached from the laws of physics, geomorphologists are at one with social scientists in working with general propositions of low explanatory power. Perhaps this is why the least determinate of the natural sciences is sympathetically accommodated within social sciences and flourishes in this setting.

Process and form

INTRODUCTION

Until 1960 it was widely accepted that geomorphology was essentially the study of landsurface forms. The nature of the processes concerned was inferred from the forms they were believed to have created. For example, most investigators adopting Matthes's 1900 concept of 'nivation' have considered this a major process in the development of periglacial and glacial landsurfaces. Landforms ranging from miniature hollows to composite cirques have been attributed to nivation, yet when the mechanisms involved were scarcely scrutinized (Thorn, 1976). Further, it seems that a morphological classification of snow patches and nivation hollows adds little in identifying the processes in such environments. It has therefore been generally and rapidly accepted that, with only a rudimentary knowledge of processes, it is too optimistic to assume that the nature of processes can be inferred from landforms. Circular arguments have been detected creeping round the literature, using the evidence of forms to support an hypothesis of their genesis and then adopting this hypothesis to explain the same forms. The response to these realizations has been so immediate, energetic and emphatic that the process–form balance of emphasis in geomorphology changed within a decade, to the extent that greater future attention to form seems inevitable.

INSCRUTABILITY OF SEVERAL GEOLOGICAL PROCESSES

It is impossible to observe some geological processes directly. This has always been the case for the study of plutonic igneous complexes. Currently, many important geological questions can be solved by utilizing the plate tectonic model, yet 'the more fundamental problem of causal mechanism must await further knowledge of deep earth structure and composition before a really satisfactory explanation can be sought' (Hallam, 1971). When igneous processes do emerge at the surface, the risks of direct observation are deadly. For example, in the ejection of glowing avalanches, the nature of underflow and the role of gas must be inferred from details of debris particle size and shape, and from avalanche morphology and bedding.

Geology is unique in that only the results of natural processes are available for inspection in many instances. Therefore, it is unsurprising that a strong tradition of inferring the nature of processes has grown up in many branches of geology and that, in consequence, the detailed study of contemporary geomorphological processes has tended to be overlooked. Today, however, geomorphologists in geological realms are by no means alone in their observation and measurement of present-day processes. For instance, in addition to the developments in studies of process dynamics in sedimentology, palaeontologists increasingly study living foraminifera.

THE SURGE IN STUDIES OF GEOMORPHOLOGICAL PROCESSES

Information on processes has emerged from all directions, quite often the product of considerable ingenuity. For example, shear stress and normal pressure changes at the contact between rock and the overriding glacier's sole have been measured by transducers set into holes drilled in the bedrock (Boulton *el al.*, 1979). Not only has the significance of layers or pockets of water in facilitating rapid ice movement been established, but also the possibility that variations in the solute content affect the motion of temperate glaciers (Hallet, Lorrain and Souchez, 1978). In the case of nivation, water-tracing observations in the Canadian Arctic indicate that the rise of the permafrost table to the ground surface underneath perennial and late-lying snow patches is critical in bringing subsoil drainage water to the surface. This water drains from the active layer of the adjacent area, much larger than the snow patch itself, initiates rillwash, and saturates the ground downslope from 'nivation' forms (Ballantyne, 1978). Similarly, many studies have been directed towards identifying rainfall parameters which strongly affect soil erosion (Moore, 1979). In stream channels, the way in which the extent of the network of actual flow varies with rainfall has been mapped (Day, 1978) and the critical role of bank collapse in channel migration plotted (Hooke, 1979). Offshore, the effect of tidal regime, bottom currents, and water circulation has been investigated in bays like Western Port in Victoria, Australia (Sternberg and Marsden, 1979).

Chemical weathering is now understood as a significant influence on land-surface formation, and studies from a widening range of lithologies have become as common as the early work on limestones. Knowledge of silicate mineral breakdown increases rapidly and mineral-organic matter reactions are being scrutinized (Waylen, 1979). Also, the geographical range of appreciable chemical weathering is increasingly recognized, being much more important at snow-patch sites than has been traditionally recognized, attributable to seasonally high temperatures at bedrock and colluvial surfaces when meltwater flows (Thorn and Hall, 1980). Soil-water residence times, critical to studies of rock decomposition processes, have been estimated by fluorescent dyes and by Ca:Mg ratios in calcareous soil waters. Such ratios can be interpreted in terms of short-residence times for high calcium values, mixing with longer-residence water with a high magnesium content (Trudgill, Laidlaw and Smart, 1980). Detailed investigations specify prodigious earth-moving in areas of intense animal activity (Williams, 1978; Hazelhoff *et al.*, 1981). Not least has been a willingness of specialists to adapt and employ their expertise in studying processes in a contrasted environment. For instance, some detailed information on tropical geomorphological processes has been gained by the recent translocation of Scandinavian experience to tropical areas.

UNCERTAINTIES IN ESTIMATING THE NATURE AND EFFECTIVENESS OF GEOMORPHOLOGICAL PROCESSES

Geomorphological processes are not easily specified exactly, for several reasons. First, many significant processes are not necessarily either conspicuous nor even visible, and the most important process in a given locality is not necessarily indentifiable prior to detailed study. Secondly, many processes operate so slowly compared with the time-span at the investigator's disposal. For example, one criterion in C. F. S. Sharpe's definition of soil creep is that it may be imperceptible except to measurements of long duration. Thirdly, there is the multiplicity of natural processes which makes the detailed nature of any one difficult to identify. On beaches, for instance, there are four contrasting hydrodynamic zones which are overlapping and compressed into a relatively narrow strip between the backshore and the nearshore (King, 1959). Fourthly, many processes are difficult, if not dangerous, to observe, with risk increasing at times when mechanical processes are most active or biochemical activity most virulent. A simple and observable process like cut-and-fill in the zone of breaking waves is not well documented due to inaccessibility even at low water. In consequence, it has not been possible to support, question, or reject the notion that shore-platform erosion occurs chiefly in the storm-wave surf-zone because the effectiveness of erosion seaward from the breaker line is not clearly established.

Even if geomorphological processes are comprehended, this knowledge is not easily integrated into interpretations of landforms. This task is most apparent for 'homologies' of landforms, the cases where different processes appear to produce similar forms known as 'convergence' to European workers and 'equifinality' in the United States. For example, small circular weathering pits occur in varied environments and, on a larger scale, rounded ponds in formerly glaciated areas might be either kettle holes or ancient pingos. Rounded summits with tor-like eminences also occur in varied environments, so that those near Fairbanks in Alaska are reminiscent of Dartmoor tors. These in turn have been likened to 'castle kopjes' in southern Africa. In coastal areas there appears to be more than one genetic type of barrier island.

Most uncertainties arise from inevitably incomplete knowledge of the fourth dimension. Above all, the representativeness of present-day processes is uncertain due to the advent and enlargement of technologies which have accelerated processes and swung the scales of earlier balances. Even where a given form is plausibly the result of a known natural process, the present intensity of that process might not represent slightly different conditions in the past. For example, the prevailing climate where periglacial microforms are actively evolving may differ slightly from that of a century ago. Similarly, slight changes in past sealevels might explain the size of the solutional nip in limestones which indent cliffs on tropical coasts just above high-water marks. Indeed, explanations of any shore platforms in terms of present agents visibly at work could be misleading. In streams, sediment yield in headwaters may resemble contemporary denudation rates, whereas tributaries entering larger rivers may have yields reflecting the rate at which debris from a preceding weathering regime is entering the channel, such as glacial deposits (Ledger, Lovell and Cuttle, 1980). Ultimately, knowledge of present-day processes may not incorporate experience of a rare but critical event. Thus, some characteristic river-valley forms can be relicts of infrequent floods and not the result of day-to-day or year-to-year conditions of normal flow.

In all situations there is the problem of deciding on a sequence of events. The process may essentially precede the form and thereby determine its character, the converse, or the two may coexist in a closely interdependent state of 'dynamic equilibrium'. For instance, in coastal depositional landforms there is such a close coupling between process and form that any distinction between cause and effect is not readily apparent (Wright and Thom, 1977). On the other hand, many workers identify too many changes in the last couple of million years for the last suggestion to be no more than a hypothetical abstraction for solid-rock landforms. A sequence of antecedent and subsequent conditions or the possibility of balance depends largely on the time-scale involved (Schumm and Lichty, 1965). Thus, in the short term in river bends, the meander is an anterior condition and helicoidal flow exists because of the meander. The meander itself, however, if considered in a broader time context, may exist partly because of different antecedent conditions of flow. The span of such time-scales depends on the relative ease with which materials are moved.

SEDIMENTS, SOILS AND GEOMORPHOLOGICAL PROCESSES

If it is unwise to infer processes from forms, conversely any anticipation that forms can be predicted from processes may be no sounder. Therefore, the study of sediments and soils has been incorporated into landform studies as these may provide significant links between form and process. Sediments and soils result from the same processes which affect landforms and reflect the rock type of the landform. Both may accelerate or retard the rate at which processes operate and influence their modes of operation, depending largely on the physical properties, as expressed by soil mechanics theory and techniques (Whalley, 1976). The study of sediments involves not only depositional structures like gravel lenses, but also the measurement of statistical parameters describing characteristics of groups of individual particles. For example, the dynamics of the swash zone on a beach, although difficult to observe in action, are reflected in grain-size distribution across the foreshore slope. In turn, grain size influences infiltration rates of the swash.

Soil and sediment analysis is particularly illuminating in areas of decomposed granite, such as the Dartmoor region of south-west England. Initially, deep weathering was attributed to humid tropical weathering, but this view was not supported by soil analysis, which showed that the eye-catching tors had been exhumed from a sandy and not a clayey weathering zone (Eden and Green, 1971). Indeed, closer examination of the quartz particles shows signs of mechanical breakage only and, in general, very little evidence of intensive chemical weathering (Doornkamp, 1974).

Sediments as an intermediary between process and form do not, however, necessarily elucidate the direction of cause and effect. In the case of pediments in the Mojave Desert, Cooke and Reeves (1972) conclude that the association of the largest-sized particles with slope angle is due to these particles representing the stable fraction of the debris at any given position on a slope. Similarly, Akagi (1980), in studying the Sonoran Desert, concludes that debris is controlled by the inclination of the slopes. These conclusions are at variance with the notions of debris-controlled slopes, initially postulated by W. Penck and Kirk Bryan. In some cases no relationship between soil and slope form or steepness is discernible. For example, in the Johor area of West Malaysia the main soil change observed, that of downslope decrease in the clay-silt content, is unrelated to slope form or angle (Swan, 1970).

In situations where mechanical action is much reduced, as in soils and some sediments, particles which have spent some time in the weathering profile may exhibit chemical changes which reflect the importance of process, climate and time. For instance, soil mapping and associated laboratory studies on samples from various parts of the Chalk outcrop in southern England illustrate the importance of continued subaerial weathering during the Quaternary in the evolution of the landsurface (Catt and Hodgson, 1976). Thus, just as the usefulness of a study of present-day processes extends to interpretations of the past, so also can the study of modern sediments, by revealing characteristic properties developed in distinctive environments, cast light on the nature of past environments decipherable from the characteristics of fossil sediments and soils. However, it must be stressed that the operation of more than one process in a given depositional environment complicates the interpretation of sediment-process interrelationships. This is clearly the case where fluvial and coastal processes intermingle, with further possible causal mechanisms such as floating ice shelves and subglacial flow to be considered in higher latitudes (Ashwell, 1975).

The degree to which sediments, soils and geomorphological processes are an integrated phenomenon is due to their shared position at the earth's land-surface, where the lithosphere is exposed to the energies of the atmosphere, hydrosphere and biosphere. Of the links between sediments, soil and process with landform, those of soils receive increased emphasis. Compared with the geographically localized areas in which sediment accumulates, or the temporal fluctuations in present-day processes, soils are nearly continuous in their geographical spread, and most of their properties are more enduring than the ephemeral changes in process intensities. Indeed, soils are often the surest indicator of significant temporal change in processes. Furthermore, individual case studies are beginning to add facts to long-standing speculations about the role of soils as a controlling influence on contemporary processes as well as a long-term expression of their past actions. Such studies are few and only rarely is a spatial sequence of several soil profiles investigated in detail. The reason why a geographical approach to studies of soil–landform interrelationships proceeds slowly is soon obvious; soil properties are several, and each may change appreciably with only slight differences in soil profile depth. Such studies may intensify as key soil properties are recognized and certain depths sufficiently representative of the entire profile for geomorphological inter-pretations identified.

A FOURTH ANGLE – MICROFORMS AND GROUND-SURFACE IRREGULARITY

The study of sediments and soils gives only glimpses of the interactions between form and process. A valuable supplement is the study of microforms, which may have significance far beyond their small scale because they reflect the nature and effectiveness of contemporary processes most readily and react most quickly to any change. For example, from a detailed study of terracettes and a review of the literature, several types of disturbance are identifiable, but a common factor in the strength imparted to the ground-surface by the vegetation cover is established (Vincent and Clark, 1976). A study of earth hummocks in Cumbria reveals that, even in temperate areas, expansion of a mesh of ice crystals within superficial deposits is probably very significant (Pemberton, 1980).

In studies of karst microforms Dunkerley (1979) concludes that the effects in limestone rills (*karren*) are mainly those of limestone solubility and the mean ambient temperature, rather than rainfall intensity or hydrological character-istics of flow. In steeply inclined karren in humid tropical Malaya, however, hydrological characteristics are important, as is the amount of organic matter lodged in the grooves (Crowther, 1979). In tropical and subtropical limestone

---

Figure 9 Illustration of ground-surface roughness as a significant geomorphological parameter, as observed near the Geikie Gorge in the Fitzroy Basin of north Western Australia
A   Slope profile, showing the appearance of the rough 'giant grikeland' on the reef front of Devonian limestone, giving way to the visibly much smoother pediment.
B   Line graph of the downslope sequence of slope-angles for the consecutive 1.5 m lengths of the slope-profile survey, with arrows linking selected points to their position on the profile. The oscillation of the slope-angle plot displays visually the degree and changes in ground-surface roughness.
C   Downslope sequence of a ground-surface Roughness Index (R.I.) which calculates the degree of 'oscillation' in the values observed in the line graph ($n = 5$). The trend of the R.I. values may suggest a progressive decline in the roughness of the steeper reef-front section, and also in the much smoother surface of the pediment.
Source: Adapted from Pitty, 1974 (unpublished).

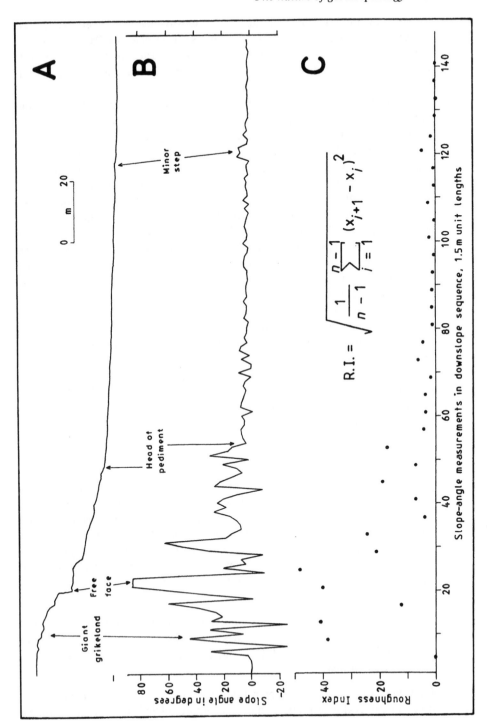

coasts a deeply cut notch or nip along the waveline is a widely recognized feature and critical to studies of recent sealevel changes (Higgins, 1980). Inland, cliff-foot recesses or tafoni may suggest moister conditions in the past, as suggested by features along the north-west margin of the Sahara (Smith, 1978).

In addition to the localized, regularly shaped microforms, a more widespread characteristic is the general condition of ground-surface roughness or irregularity. It is particularly prominent in periglacial areas on unvegetated debris slope and scree accumulations, seasonally modified by avalanche or water flow (Church, Stock and Ryder, 1979). A review of early comments prompted the conclusion that 'slope irregularity may prove to be one of the most vitally important aspects of slope character' (Pitty, 1969; 18–23). Indeed, slope roughness has been found to decrease with age on small moraines in eastern Canada (Welch, 1970) and seasonal changes in the roughness of ploughed fields specified (Reid, 1979). Figure 9 illustrates the range and geographical variability of ground-surface roughness in an example of a slope profile typical of smaller pediments in the area of northern Western Australia, described in detail by Jennings and Sweeting (1963). The profile (Fig. 8A) shows the visual contrast between the rough surface of reef limestones above the smoother pediment and since measurements were made in 1.5 m consecutive steps, the line graph (Fig. 8B) emphasizes the contrast diagramatically. Figure 8C shows how a 'roughness index' can specify the range and geographical trend in this parameter and how these could be suggestive of the processes at work. For example, the very smooth lowest section downslope from the 'minor step' may reflect planation by the Fitzroy river at peak flow during monsoonal rains.

BASIC SCIENCES, GEOGRAPHY AND GEOMORPHOLOGICAL PROCESSES

The enlargement of landform study to include processes, sediments and soils raises the question of geomorphology's outer limits. There is the risk not only of venturing beyond what is strictly required for landform interpretation, but also, in consequence, claiming for, or handing to, the geomorphologist, studies which rest more appropriately in other hands. Forty years ago Bagnold (1941) anticipated that the more detailed investigation of processes would become an integral part of landform study. He surmized that the geomorphologist would not be content with

> study for its own sake of the shape and movement of sand accumulations, until he knows *why* sand collects in dunes at all, instead of scattering evenly over the land as do fine grains of dust, and *how* dunes assume and maintain their own especial shapes.

However, since he also indicated that 'the subject of sand movement lies far more in the realm of physics than of geomorphology', the risk of venturing beyond a brief was also posted early. Similarly, in sediment transport in rivers, hydrodynamicists and hydraulicists study energy relationships between particles and the transporting fluid, whereas the geomorphologist is more concerned with the more tangible amounts and rates of material moved and the location of source and redepositional areas. A limit is also recognizable in glaciology since glacier flow, being one of the most useful illustrations of plastic flow of a crystalline material, is studied intensively well beyond the geomorphologist's bounds. In fact, the particular process causing folds in ice may also operate in rocks. Thus, the physics of ice movement, apart from internal mechanics of glacier motions, is equally relevant to some other branches of earth science because of the kinematic analogy with recumbent structures in orogenic belts. Limits are also suggested where exact scientific treatment would be an oversimplification. Commonly, these occur where several variables interact simultaneously, are due to the incompleteness of evidence and, most fundamentally, arise because of continual drifts in the intensity of processes over periods of time.

Greater insights into the nature of geomorphological processes are gained from groundings in basic chemistry, physics and biology. However, insistence on this as an indispensable formal background for clear and easy understanding is, instead, a source of confusion and stress where geomorphology is thought, taught and learnt as part of a geography within a social-science framework. Any such insistence may also be imperceptive. Compared with the earlier emphases in geomorphology laid by the European chronologists, led by Baulig or by Wooldridge, or the later 'landform geography' in the United States summarized by Zakrzewska (1967), it is a process-orientated geomorphology that brings most to bear on human geography (Robinson, 1963). Phenomena such as soil creep, surf or channel shift have much more bearing on the everyday moods and matters of the human condition than U-shaped valleys or relict Tertiary landsurfaces. Thus, the contemporary, process-orientated geomorphology can move as purposefully and readily towards the social sciences as to the basic natural sciences.

Qualitative and quantitative aspects

INTRODUCTION

The complexity and irregularity of forms and processes, together with complicated changes with time, promote an essentially qualitative approach as the only realistic way of approaching many geomorphological problems. Also, any antithesis between qualitative and quantitative aspects is largely false, due to their essential interdependence. However, since 1945 sufficient regularity has been recognized in some forms and several processes to encourage more exact approaches. In other instances qualitative approaches have proceeded as far as they can go. For instance, the many studies of the late Pleistocene history of the

Hudson River estuary, before the advent of radiocarbon dating, were 'inconclusive, incomplete and misleading' (Weiss, 1974). The clarification brought to such time-scale problems by the measuring of recent geological time, insoluble in terms of earlier vague controls, encourages searches for more exact methods on all fronts (Goudie, 1981).

The quantitative element in geomorphology differs depending on whether time, process or form is considered. In the measurement of time many disciplines such as archaeology, botany and geochemistry provide techniques and experience in their use. In process studies long-established familiarity with quantitative techniques already exists in specialisms such as hydrology, hydraulics, agricultural engineering, soil chemistry or sedimentology. In contrast, the quantification of landform is the particular responsibility of the geomorphologist. Consequently, with little experience available in related subjects, quantitative studies of form are more rudimentary, and those of the functional interrelationships between form and process very few. It is striking that contemporary geomorphology's strengths are ringed around its multi-disciplinary periphery rather than congregated at its formal core.

DATA

The value of qualitative reasoning behind the collection of data is one of the most important aspects of quantification. Many concepts require clarification before they are represented by numbers rather than in retrospect, and the importance of design before rather than after measuring cannot be over-stressed. This is evident in four aspects of handling a large number of measurements.

First, in defining a parameter representative of the unit or property to be measured, convenient measures will inevitably be partly arbitrary, a point already demonstrated in considering maps as a source of descriptive measurements. In sedimentological studies only the effects of grain size are usually considered, since particle shape is difficult to quantify. Similarly, the erodibility of soils is difficult to express in single numbers. Defining a parameter to represent the scale or a property of static forms or sediments is difficult enough, but in the moving phenomena of process and sediment studies, some changes elude definition by being too slow or too rapid.

Secondly, in collecting data, expectations of accuracy are linked with the intrinsic variability of the phenomena. Also, for some variables, such as the cohesiveness of a sediment in relation to its resistance to erosion, no field method has been developed. Ceaseless motion, as on the shore, makes it difficult to measure sea state, to distinguish edge waves from the orbital velocities of the incident waves, or to collect sediment samples in the surf. Sophisticated apparatus requires precautions and even restrictions in its use. Equipment and markers employed in continuous measurements may be accidently disturbed. A geographical approach to some problems may be restricted to a small number of sites if bulky or expensive equipment is involved. Ideally, many measurements are required to encompass both the complexity of environmental interrelationships and also the range of geographical variability. For instance, channel cross-sections may differ considerably within even a short reach, and trends in at-a-station hydraulic geometries are only poorly defined by less than ten cross-sections. They may change annually or during bankfull floods (Richards, 1977).

Thirdly, data already collected for some other purpose can be redeployed for geomorphological enquiry. The use of maps as a source of indices to describe landform and stream pattern is well known. Also, huge volumes of data collected by water engineers interests fluvial geomorphologists, and nearly all process geomorphology makes use of climatic data. The limitation is that the scale, siting, quantity and measurement units are not always ideally suited to a geomorphological enquiry.

Finally, the statistical character of geomorphological data is increasingly comprehended. Its worth is confirmed by consistencies revealed in the data collected over the last twenty years. Also the phase is passing in which absence of *a priori* knowledge of the relative importance of features meant that as many as possible were measured. More is now confidently known about the minimum precision needed in a particular study. On the one hand, superfluous accuracy which serves only to buffer the investigator against ill-founded criticism can be identified. Indeed, qualitative assessments of such characteristics as shape, hardness, colour and texture can be made readily and inexpensively, whereas the setting up of precise measuring apparatus may be time-consuming, costly and, in terms of capacity to differentiate among objects, unnecessary (Andrews and Estabrook, 1971). The ultimate economy is the presence-or-absence attribute (binary or two-state nominal data) which, in view of the wide elements of uncertainty in many landform interpretations, is adequate for many geomorphological investigations. Conversely, the ordinal-scale representation of stream ordering is seen increasingly to be too crude for summarizing hydrological and landsurface properties of drainage basins.

CORRELATION

Before calculating the degree of association between two geomorphological variables, the statistical distribution of measurements within a sample must be scrutinized. Measurements may bunch near one end of their range. For instance, the smallest of a drumlin field may be only half the size of the majority, which the largest may exceed by several times. In some cases, a widely employed measuring range is too narrow to include extremes, the particle sizes of boulder clays being an example. There is also the 'straggler' or statistical 'outlier', the isolated atypical case which, having an extremely high or low value, may have a strong but misleading influence on calculations as a

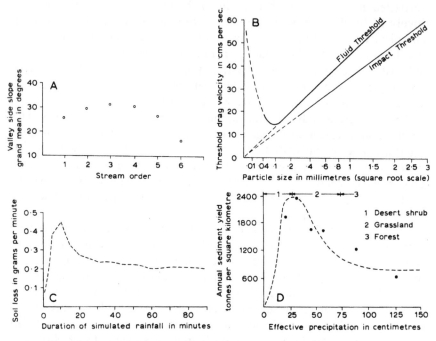

Figure 10 Some non-monotonic relationships in landform studies
A   Variation of mean valley-side slope angles with stream order, for the Lighthouse Hollow drainage network, Poquonock, Connecticut.
B   Variation of the threshold drag velocity in air with particle size.
C   Soil loss in runoff during a 90-minute simulated rainfall application. The soil is a silt loam from Maine.
D   Variation in sediment yield with varying amounts of effective precipitation. Wilson (1973, *Am. Jour. Sci.*, 273, 335–49) argues that this 'Langbein-Schumm Rule' is based on limited data and is not applicable on a worldwide basis. More copious data suggest a bimodal curve, with peaks at approximately 750 mm and 1750 mm mean annual precipitation, which compounds difficulties in analysing sediment yields statistically.
Source: Adapted from A Carter and Chorley, 1961; B Bagnold, 1941; C Epstein and Grant, 1967; D Langbein and Schumm, 1958.

whole. Such a case is the River Ganges at Patna where the gradient is unusually low for a large braided stream (Osterkamp, 1978). Perhaps the most commonly encountered geomorphological 'straggler' is the steep angles of free faces in sets of slope-angle data (Fig. 9B). Many phenomena may have more than one dominant size class in a frequency distribution. This characteristic, however, has not deterred the calculation of statistical measures such as skewness and kurtosis in sediments, since there is no satisfactory way of dealing with polymodality. In slope-angle data, polymodality may be resolved by noting that different subpopulations may occur on different portions of the slope (Cooper, 1980).

The correlation between two geomorphological phenomena may not plot as a straight line. Behaviour may even change suddenly when a certain threshold value is exceeded or with a new factor emerging at a particular time or beyond a certain distance (Pitty, 1971). In several circumstances a trend in *y* values may be reversed beyond certain values of *x* (Fig. 10). Rates of change may also be non-monotonic. For instance, in eastern British Columbia an initially slow migration of channel bend reaches a maximum rate as the radius/width ratio approaches 3.0, and then rapidly declines (Hicken and Nanson, 1975). A particular type of bivariate regression, in which the relative rates of change of two shape variables is demonstrated, is described by some workers with the metaphorical term 'allometry'.

Even if data have no unusual features, the complexities of the natural environment may make a simple and excellent correlation unlikely. For instance, fluvial data suggest a pool-riffle spacing equal to half the straight-line meander wavelength, but many exceptions exist, such as where several pools are found on a large meander bend. Further, pools and riffles are well developed in some straight reaches, and the regular spacing of pools can thus be independent of channel pattern. Finally, correlation analysis may indicate no more than that other types of evidence are required. Factors widely assumed to be closely related, as in pediment studies, may reveal no consistent relationship on analysis.

The interpretation of correlations is never easy, sometimes not obvious, and occasionally apparently the reverse of what might be anticipated. This possibility is illustrated by rock-creep studies in Colorado. Although increased rock creep at times of higher precipitation might be anticipated, attributable to a lubricating effect of moisture, measured creep rates were greater during drier spells (Tamburi, 1974). The explanation offered in this particular case was that diurnal temperature range controls rock expansion: rock creep is greatest due to radiative atmospheric heating and cooling under clear skies, and least beneath cloud cover when rain may fall.

EQUATIONS

Causal laws can apply only when the connection between phenomena is invariably the case. This deterministic precision of invariable mathematical relationships therefore rarely exists in geomorphology. Instead, some individual element is allowed in each case and, in such stochastic laws, probability only is implied and replaces necessity. Nonetheless, correlation and regression equations are often as far as generalization can proceed without introducing extravagant simplifying assumptions. Mathematical relationships cannot allow realistically for changes in intensity with time, such as climatic changes in the Pleistocene or episodic pulses over shorter, scattered intervals. Not least there is the 'kaleidoscope of hydrologic conditions caused by the heterogeneities of geology' (LeGrand, 1962).

Four types of equation, in addition to regressions between two or more variables, are encountered in contemporary geomorphological literature. First, R. E. Horton's work on the geometrical properties of drainage basins has been extended to demonstrate an invariability in these properties which invites mathematical treatment, mainly by A. N. Strahler and his associates. However, the regularity revealed in a Hortonian analysis is successful only in certain areas (Eyles, 1974). Also, some of the 'laws' are statistical probability functions applicable to any branching system and are not derived from the landsurface but follow automatically simply as a consequence of the manner in which stream order is defined (Pitty, 1971).

Secondly, an analogy between a ground-surface shape and a mathematically defined shape is occasionally drawn. Such analogies date back to the delta, and include Galileo's suggestion that the long profile of streams approaches a quarter circle. Workers who have attempted to use empirically fitted curves to reconstruct extensions of former long profiles were soon hemmed in by uncertainties. The inevitability of irregularities in most profiles, due to resistant rock outcrops, means that at best a profile is segmented into a series of curves which would each require separate equations. The alternative is gross oversimplification, as attempted by Devdariani (1967) who could only fit an equation to stream profiles by eliminating breaks and irregularities in order to assume a smooth profile, ignoring three-dimensional complications such as meandering, and by assuming an unchanging climate over time and homogeneous lithology. Similar restrictions apply to smooth coastlines or to most hillslopes. An agricultural scientist has considered briefly the application of three-dimensional formulae to ground-surface shape (Troeh, 1965). He concludes that the complexity of a series of hills and valleys is so great that it would be impossible to describe them even approximately with a three-dimensional equation. Even if derived, it would be difficult to assign meaning to the coefficients and items of real interest, such as slope gradient, would not be readily obtained from them.

A third type of formula, the theoretical mathematical model, has recurred in the literature for at least a century. These equations start with the untested assumption that landsurface shapes and the processes operating on them should be definable in terms of the laws of physics. The most general are the continuity equation, which states that matter is not lost, and the equation of conservation which states that energy can neither be created nor destroyed; 'they alone tell us nothing about the surface form of the landscape' (Leopold and Langbein, 1962). The equations used to formulate such speculations should be clearly distinguished from those which summarize large amounts of actual observations. Their values are not measurements but assumptions and predictions. In fact, discussion of such equations may give no clue on how certain variables might be determined or even defined. This applies, for instance, to the length and depth of the 'equilibrium bottom profile' in Bruun's formula to predict shoreline recession with rising water level. Amidst these equations, one often finds a plethora of mixed metaphors, such as streams with 'heads of unbranched fingertips' which might behave like 'a sluggish governor on the old steam engine'. Such approaches usually depend on basic assumptions unacceptably oversimplified for workers with real-world preferences. Also, landform development involves as much chemistry as physics, and theoretical notions about ground-surface shapes and associated processes, conceived and nurtured in exact and laboratory-controlled sciences, invariably underestimate biological factors. The homogeneity of geology is a near universal but questionable assumption in the mathematical model. It is therefore particularly significant for the progress of the subject that the author of the pioneer, mathematically based *Theoretical Geomorphology* (Scheidegger, 1961) has recently amassed data demonstrating that a preferred orientation of valleys in Switzerland corresponds quite well with joint orientations (Scheidegger, 1979).

Fourthly, formulae devised and employed by technologists may appear in geomorphological literature. Their purpose is to predict the range of natural processes and properties likely to be encountered where engineering installations are planned. Because of the wide safety margins, approximate predictions are adequate and such formulae, based partly on measurements and partly on theoretical considerations, are essentially a substitute for measurements rather than a summary. Such formulae can help the geomorphologist planning to collect a series of measurements, as they can indicate approximately the range of values which methods and equipment will have to encompass. One short-cut device used in deriving empirical formulae for technological purposes is to use a given set of measurements in estimating mo. e than one parameter in the equation. If such parameters occur on both sides of an equation, their degree of association is artificially exaggerated. Another practice is the adoption of a particular formula, such as a power function, when there is no reason to assume that such a function necessarily represents the relationship between the variables.

QUALITIES IN LANDFORM STUDY

In addition to the trend towards greater quantification, there are also indelible qualities in landform study, consistent with the hypothesis that the subject shares an equally low level of explanation with the social sciences. Most apparent is the natural beauty of scenery. Works of scholarship survey the history of the subject against the backcloths of grandeur where pioneers once strode (Chorley, Dunn and Beckinsale, 1964), or balance the emotional arguments of the early days (Davies, 1969). Also important are collations of early statements on recurrently popular topics, since the observations and insights of pioneers can sometimes repay reconsideration in the light of further evidence. The classic examples are now continental drifters A. Wegener and A. du Toit, but a more strictly geomorphological example is a review of early explanations of granite-boulder forms (Twidale, 1978). A growing trend has been the overlap with historical geography and the great skill shown in using historical documents to chronicle geomorphological changes during recent decades and centuries. For example, Norwegian glaciers expanded rapidly in the late seventeenth century, reached advanced positions in the mid-eighteenth century, and then began to wane. During this period of the 'Little Ice Age', land-rent assessments known as the *landskyld* provide detailed information about the incidence of landslides, rockfalls and avalanches. From these records, Grove (1972) establishes that the years 1687, 1693 and 1702 were particularly prone to natural disasters.

CONCLUSIONS

A feature of post-Davisian geomorphology has been greater attention to quantities, and a current strength of the subject is the wealth of detailed measurements that has gradually accumulated. This emphasis, however, has not always yielded positive results. There has been some 'wildcat quanti-fication', with measurements made blindly in the hope of striking some reservoir rich in significant correlations. Several equations have been put forward without supporting data, despite the early and authoritative suggestion that a theorist is identifiable with a failure to test hypotheses adequately (Gilbert, 1886). The difficulty of encompassing geomorphological phenomena entirely by quantities and equations is evident in the huge cost of hydraulic-engineering scale models of dynamic landscapes and in the trial-and-error procedures involved in their study. As a compensation for limits on the degree to which quantification can be successfully applied, the qualitative realms of library shelves, old documents and charts, and the aesthetic qualities of scenery itself open up as attractive directions in which geomorphological evidence can also be found.

Laboratory and field work

THE FIELD EXPERIENCE

If something as legitimate and useful and geographic as exploration has been purged from geography, what else has a chance? (Bunge, 1973)

The first geomorphologists were explorers or the field geologists preparing maps on reconnaissance surveys. A wide geographical range of field experience remains an asset, either for generalizing or for heightening appreciation of the distinctiveness of a given area. Thus, Ager (1970) suggests that it is 'necessary to go and see for oneself before one realizes that the limestone of the Mammoth Cave of Kentucky is comparable to that of the Cheddar Caves of Somerset'. The mere fact of being in the field increases the chances of discovering crucial evidence, recognizing a particularly instructive locality, or witnessing an infrequent event. For instance, many observations were made of the Mt Huascaran landslide in north-central Peru, which moved some 2 million m³ of rock in May 1970, because several scientists chanced to be in the area at the time. A classic example is Buckland's observations of a landslide in Devon in 1839 since a professional artist was also on hand to depict the results (Thornes and Brunsden, 1977; 52).

Since 1945 the emphasis has passed from exploration and extensive field travel to more detailed repetitive work. For example, any field measurements of river-bank strength need to be taken at bankfull or near-bankfull stage because of the control of bulk moisture content on soil strength (Park, 1978). Thus, and typically the case in many contemporary process studies, repeated visits to the field site are needed to obtain a sufficient range of data from which critical or threshold levels can be predicted. However, some physical challenge persists because significant geomorphological changes usually occur at times when physical risk is greatest, such as bankfull stage! Equally, the best time to inspect material from an eroding cliff is in late winter after a severe storm. This, ideally, would be the critical time for collecting samples in the zone of breaking waves, as storms highly intensify shoreline processes, yet residual waves from such storms may remain high. Thus, an attempt to investigate the sea-bed in Repulse Bay, Hong Kong, during the aftermath of Typhoon Viola in September 1969 had to be abandoned because it was impossible for divers to remain stationary and sea-state conditions were dangerous (Williams, Grant and Leatherman, 1977). Irregular working hours may be necessary: for instance, floods may peak at any time of the day or night. In excavating ice-cored moraines, Østrem found that pit-digging had to be done partly at night because, with the sun shining on the pit walls, melted masses slumped down. In hot deserts, time for surveying is limited by the rapid development of heat haze. The discomforts of extremes of temperature, rain and wind, sand in the eyes and flies in ears and nose, can all make accuracy in calibration and measurement a vexing task for those not inured to exposure to the elements of the environment. Thus, despite its glinting hardware, the increasingly scientific character of geomorphology maintains a demand for some of the enduring qualities and mental fibres of the traditional explorers of the past. Unfortunately, leading exponents are often glaciologists, studying areas remote from man. Such directions can be seen as contrary to those of a man-centred geography, yet human geographers may yet find, beyond the span of a particular three-score years and ten, that mankind's greatest challenge is the second coming of the flood if ice-caps and glaciers continue to melt.

In education the role of field work is traditionally stressed, sometimes weakened by excessive emphasis. Geology 'At its roots is field-orientated and may include a field course in the educational curriculum. . . . If we have lost, in some measure, direct contact with the earth, this may lead to a loss of curiosity and creativity' (Steinker, 1979). Geographers often reminisce on the broadest advantages of field work. Thus, Harris (1979) recalls his early training in which 'The field work was not only a great learning experience; it was also a time for camaraderie and sustained discussion of geography and just about everything else'. Clearly, old field workers never die; they just fade away.

LABORATORY ANALYSIS

The sheer volume of field measurements and samples increasingly confines the geomorphologist to the laboratory. Field measurements are fed directly into some style of statistical analysis and diagrammatic representation. In the laboratory examination of materials, however, differences from field conditions must be recognized. Certain analyses where properties such as the proportion of clay in a soil or the estimation of the amount of sediment in a water sample are concerned, involve only sampling errors in any difference between the field condition and the laboratory test. However, the laboratory analysis may not specify the property which is most critical in field behaviour. For instance, the three diameters of pebbles are the most critical factors accounting for sorting on Chesil Beach. However, the least critical is the intermediate diameter, yet this is the axis which is most closely related to sieve sizes used in particle size analysis (Gleason, Blackley and Carr, 1975). Another category involves properties which may change during transit to the laboratory, such as soil moisture or pH. Here distance may be the criterion by which a choice is made between rapid transport of samples to the laboratory or the taking of bulky or sensitive apparatus into the field.

LABORATORY EXPERIMENTS

One of the most incisive investigative tools in science is the experiment in which the condition under which observed phenomena are held is controlled. Laboratory experiments, involving the dynamic interaction of artificially controlled parameters, must be clearly distinguished from the laboratory analysis of properties. Such experiments, like soil shear strength, permeability or freeze-thaw cycles, may illuminate the mechanisms of natural processes, but do not have an exact bearing on natural relationships. For chemical reactions, the essential difference is the period of time over which natural processes operate. Materials which show negligible solution under controlled laboratory conditions may, over long geological periods, be eliminated from a weathering profile. Another major difference is the sterility under which most laboratory experiments are conducted, compared with the natural environment teeming with life and its biological and biochemical processes.

A recent trend to broaden the scope and representativeness of laboratory-style investigations, evident in soil science as well as geomorphology, is to install equipment or emplace test materials under controlled conditions in the field. An example of such emplacements in the natural environment is the burial in soil of rock discs of known weight. In one study, weight losses at humid tropical sites were compared with those obtained in humid temperature environments, revealing losses about 3.5 times greater in the humid tropics (Day, Leigh and Young, 1980).

SCALED MODELS AND SIMULATIONS

With expansion of laboratory facilities and capacities, computer simulations and scaled models have been devised. Either approach facilitates the Baconian procedure of studying one factor at a time whilst keeping all others constant or controlled. Such regulation is impossible in field conditions where several variables may vary simultaneously and some continuously. Scaled geomorphological models have been of particular interest to American geologists in the 1970s. However, despite the attractive size of laboratory scaled models, and contrary to G. K. Gilbert's belief and one held by Francis Bacon himself, artificialities do limit the applicability of any findings to natural situations (Pitty, 1979). Not all dimensions can be scaled down. For instance, smaller silt-sized particles behave in a different way from sands. Clays are not used because flume distances are too short for clay-sized particles to settle. The depth of the model has to be exaggerated sufficiently to make turbulent flow possible. Erosion in a flume or tank of randomly packed sieve fractions may not resemble that of an exactingly selected and packed natural sediment. The use of an artificially uniform particle size is also a limitation. For instance, the amount of energy generated by a vortex is related to particle size. Only when coarse particles are mixed with fine do the larger vortices form around them which would have sufficient energy to lift finer particles.

Commonly, the medium in scaled models is incoherent, whereas much of the landsurface is founded on solid rock. Thus no scale-model experiments of the transformation of rocky coasts in response to wave action has been conducted in contrast to many such studies of sandy-beach changes. The difficulty is one of choosing a material which has mechanical properties similar to those of natural cliffs (Sunamura, 1975). Conversely, such erosional processes, like the incision of a river into bedrock, cannot be analysed adequately in the field due to the spans of time involved for appreciable change to occur (Shepherd and Schumm, 1974).

CONCLUSIONS

The ultimate value of any geographical concept can only be found when it is confronted with the conditions in particular places. (James and Mather, 1977)

Field experience incorporates the unsimulated advantage of contemplating reality and renewed emphasis on field data is anticipated during the 1980s

(Graf *et al.*, 1980). The risk may be when 'the geologist seems to delight in local complexities. He takes pleasure in, or overemphasizes, the exceptions' (Ager, 1970). Extension of interpretations of laboratory experiments into the context of natural conditions is hampered by doubts on the closeness between the simulation and the natural conditions. Flumes, for instance, yield higher ratios of plan dimensions to discharge than occur in natural streams (Dury, 1976). Agricultural engineering experiments are not readily translated to field situations, partly due to the use of simulated rainfall (Imeson, Vis and de Water, 1981). Only cautiously qualitative conclusions can be drawn, and these are usually applicable within narrow limits only. Any more precise interpretation depends on further measurements made at suitable field sites. Certain complexities of the natural situation may escape the attention of the laboratory experimenter. For instance, it has been assumed that ripple marks are orientated perpendicularly to oncoming winds. Field observations reveal that, on barchan dunes, the downwind perpendicular to the ripple strike is deflected by as much as 35 degrees (Howard, 1977). Conversely, laboratory experiments may indicate features not previously recognized as significant in the natural environment. For instance, bar systems were recognized as a major feature of inshore zones only as a result of wave-tank experiments.

A major skill in laboratory work is an appreciation that the appropriate technique is not necessarily the most sophisticated. Thus, in studying deflation, Marion Whitney (1978) used a compressor for higher velocities or for work with larger specimens. However, for exercising precise control and for observing the behaviour of fine test materials, lung power was more instructive. Ultimately, level of specialization and personal physique, health, habit and make-up influence the striking of an individual's balance between laboratory and field work and in the adoption and adaptation of equipment. Fortunately for all interested in geomorphology, some individuals are more at ease with an eye glued to a microscope and others would not deign to use a supercharged helicopter or submersible, even if the chief technician acquired one.

Broader conclusions again suggest the minimization of dichotomies. For example, a coauthor of valued laboratory manuals stresses that 'a better *integration* of field and laboratory work is most necessary' (Pettijohn, 1956). He recommends that laboratory work be done in conjunction with field work, preferably by the same person 'and in all cases subordinate to, and dependent on the field study'. He suggests that perhaps the best advice ever given to geologists was that of Charles Lapworth, namely, 'map it, and it will all come out right'. Where both the geologists and geographers of academic spheres have withdrawn from the field frontier, they run counter to the popular interest that has grown for environmental concerns in the 1970s.

Finally, another zone of geomorphology's juxtaposition with the social sciences is clearly discernible. The difficulty or inapplicability of controlled experiments in geomorphology resembles the situation in social sciences.

Francis Galton's eugenics was one of the first social-science ventures to demonstrate that controlled experiments involving human beings rapidly encounter ethical objections. Thus, many social sciences, like geomorphology, advance without resort to dubious controlled experiments.

The role of geomorphology

CONTRIBUTIONS TO OTHER EARTH SCIENCES

A major role of specialized landform study has been to contribute to geological knowledge in general. For instance, geomorphology might reflect the tectonic conditions which prevailed during the evolution of Africa's relief (Kennedy, 1962). However, in the last thirty years, techniques of absolute age determination using radioactive decay rates have relieved geomorphologists of the highly controversial role of attempting correlations of age using only ground-surface form or altitude as evidence. A more enduring role is at the basic level of geological mapping where lithologies between outcrops might be inferred from 'features', an intuitive appreciation of the significance of landform or ground-surface roughness. Often, the study of specific localities is aided by geomorphological evidence. The degree of dissection of glacial features may distinguish boulder clays and moraines of different ages. In tectonically active areas small fault scarps and offset drainage demarcate lines of Quaternary movements. In engineering geology geomorphological interpretation is important where subaerial displacement means that outcrops are no longer *in situ*. Thus, phenomena such as cambering may cause problems in civil engineering projects if landform history is overlooked.

Within a given area of homogeneous rock, soil types tend to form a predictable sequence or 'catena' downslope, as was first demonstrated in 1936 by G. M. Milne. This relationship and associated changes are often due largely or indirectly to slope form through its influence on moisture conditions and drainage in the soil. This relationship is a basic premise of soil mapping at a local scale, and specialist geomorphological contributions are readily integrated into soil-survey techniques (Conacher and Dalrymple, 1978). In hydrology, the natural movement of water on the ground surface and into and out of the soil depends on landform and ground-surface roughness. Thus, many of the observational features to be used in solving hydrological problems lie within the scope of geomorphology.

Geomorphology also shares with other earth sciences a common concern for reconstructing former environments. Numerous relics of the action of different climatic regimes in the past may be recognized in the forms of the landscape. For instance, cliff-foot caves in the Transvaal are evidence of former active springs in a damper climate (Marker, 1972). Associations between high lake levels and interglacial or interstadial phases in tropical regions have been clearly established (Street and Grove, 1979). For instance, in

Australia the level of many lowland lakes fell sharply after 26,000 yrs BP and deserts may have expanded (Bowler, 1976). Depositional landforms are the surer indicators of past climates, but occasionally climatic controls seem important in erosional landforms, as is most frequently observed in corrie orientation (Derbyshire and Evans, 1976). In such cases any palaeoclimatic reconstructions are elusive, due to the uncertainty of the ages of the erosional forms (Unwin, 1973). Indirectly, landforms may serve as repositories of valued pollen records. Kettle-holes and depressions in landslip areas are hollows often filled with peat and its associated pollen record, and the much rarer receptacle of a volcanic cone is much prized (Kershaw, 1976).

Examples demonstrating the integration of geomorphological findings into other earth sciences are thus readily listed. Perhaps more vital but less tangible is geomorphology's heavy responsibility in education at introductory stages. Subjects like geology and specialisms like hydrology and soils are not widely taught at school level, and students' first instructions and field demonstrations involving rocks, water and soil are commonly encountered in the geomorphology component of physical geography curricula. This encounter occurs when pupils are at their most open-minded, impressionable and emotionally capable of taking the Earth to their hearts. One of the most significant roles of landform study, therefore, may be the responsibility of initiating interest in any one of several branches of the earth sciences.

APPLIED GEOMORPHOLOGY

the new type of natural sciences wish to bring about the social control and direction of natural energy previously merely acting spontaneously or just lying dormant. In this way they are striving to help mankind to become a sage ruler of the world (in the noble sense) and a responsible director of events. (Jakucs, 1978)

Applied geomorphology describes the deliberate attempt to concentrate geomorphological expertise on the solution of practical problems, first recognizable in exercises at the beginning of the present century (Cooke and Doornkamp, 1974). Such applications of the detailed geomorphological map are now frequently stressed (Brunsden *et al.*, 1975), with the utility of maps of landslide-hazard areas being immediately apparent (Johnson, 1980). East European workers, in particular, demonstrate how these maps can indicate landsurfaces suitable or unsuitable for agriculture, communications and housing. In process studies the distinctiveness of a geomorphologist's contribution is less because here work is usually augmenting or duplicating that of established specialists (Fig. 11). Growing familiarity with process studies and interdisciplinary techniques, however, enables the geomorphologist to examine a range of practical problems, such as the disruptive effect of periglacial processes on built structures (French, 1976). More widely, it is the effect of artificial installations on natural processes that is considered. For

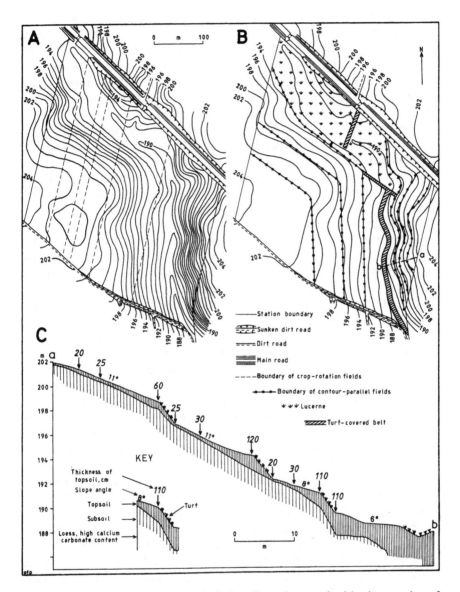

Figure 11 Application of geomorphological studies to improved cultivation practices of loess soils on the Lublin Plateau in Poland
A   Map of the Sławin Agricultural Experimental Station as it was in 1948.
B   Rearrangement of fields, changed land use, and introduction of turf-covered belts.
C   Cross-section (a–b of Fig. 11B), showing soil profiles developed over a twenty-year period after protection measures were introduced.
Source: Adapted from Ziemnicki and Repelewska-Pekolowa, 1972.

example, despite their long use in coastal engineering for trapping littoral drift or retarding beach erosion, groins introduce very complicated changes in sediment movement and beach morphology which is not readily predictable (Orme, 1980). Other examples include sedimentation in rivers, reservoirs, and irrigation and water-power plant intakes. Frequently, this sediment comes from channel-bank erosion, augmented by artificially accelerated erosion. Both are closely linked to geomorphological studies.

The concept of applied geomorphology is particularly useful in helping to define the scientific scope of landform study. Applied geomorphology overlaps substantially with adjacent subjects and much of its subject matter is equally readily identified as a component of engineering geology. Indeed, of the numerous articles reprinted in a collection of 'environmental geomorphology' (Coates, 1973), very few contributors are readily identifiable as placing geomorphology as their primary interest and even fewer are recognizable as geographers. In contrast, the practical discussion of 'terrain evaluation' (Mitchell, 1973) reminds us that the only domain to which the geomorphologist has first claim is that where applications refer to morphological characteristics only. This role is exemplified by geomorphologists' contributions to survey teams describing land use and evaluating land capacity, notably in Australia (Ollier, 1977).

The concept of applied geomorphology is less useful where, by implication, the pursuit of remaining aspects of the field are deemed less relevant. For instance, Hails (1977) suggests that 'some geomorphologists are still reticent about applying their expertise to the solutions of practical problems and prefer to pursue their individual academic interest'. However, as any educationalist knows, it is difficult to communicate interest except by example. Also, the latter part of Hails's assertion is a *non sequitur*. Those who decline to step aboard the utility bandwagon do so largely because of a persuasion that the professional role of educationalist is extremely useful. Indeed, the identification of the topics of applied geomorphology is particularly revealing in that, time and again, it is education rather than expertise that is involved in their solution, particularly within political regimes where the individual is entitled to object to planning proposals. Repeatedly, flood-control exercises reveal that the main difficulty is not one of shortage of technical expertise. The main effort is usually needed in enlightening the government authority and the public in their appreciation of the significance of scientific findings and technological potentials. Similarly, Dolan (1972) concludes that a technical breakthrough in the field of preventing shoreline erosion is unlikely. Instead, any breakthrough must be made in public policy and in an increased level of technical awareness of those responsible for shoreline management. In the case of human responses to the urban hazards of quickclays in Ottawa, a main finding was the need for a vigorous public-education programme 'with clear and forceful information in the form of maps and literature' (Parkes, Parkes and Day, 1975). In short, the first necessary ingredient of a sound environmental management recipe is an informed public (O'Riordan, 1971). This priority applies equally where 'hundreds of millions of the developing world's urban and rural poor co-exist in an increasingly precarious relationship with their natural surroundings'. Among the constraints which inhibit state action is an 'all-too-frequent indifference' (Torry, 1979).

As examples of practical problems to solve are localized, applied geomorphology tends to be highly idiographic. The enormous duplication of basic enquiries for each case can be offset only by sustained academic synthesis. In the field of conservation, for instance, there must be some evaluation of criteria for identifying what to preserve, an exercise in which there is no substitute for scholarly detachment (Krieger, 1973). With so much scope for both communicating and for collating knowledge, it seems that geomorphologists need not stand on their heads to uphold their vital utilities.

HUMAN GEOGRAPHY AND APPLIED GEOMORPHOLOGY

With only slight refocusing of perceptions, many of the practical problems with which the applied geomorphologist assists are basic subject matter for students of the 'man-and-nature' theme in human geography. The only difference is that of involvement, on the one hand, and description and communication on the other. In consequence, this coincidence of interest makes applied geomorphology a particularly valued contribution to human geography. The geographical quality is added by the regional distribution and contrasts at several scales of perceptions of practical problems, depending on political, social and religious contexts and the setting against a background ranging from affluence to poverty. This is most apparent in the sensational elements of applied geomorphology, where the planning for human occupance includes zones of natural hazard (Burton and Kates, 1964), as in winter-sports expansion in the recreation industry (Ives *et al.*, 1976).

Geography is not socially and politically neutral. Examination of ethical questions increases, which brings political overtones and undercurrents into play. A geomorphology with no explicit emphasis on application is free from political and social suppositions. An important role which geomorphology still plays, therefore, is to offer for the geographer's study the neutral subject matter of a natural science, but one with a level of explanation and method familiar to most social scientists.

3 Basic postulates

Catastrophism and Uniformitarianism

though I persuaded myself of the absolute safety of my position, I freely acknowledge that the advent of the avalanche alarmed me. (Galton, 1863)

INTRODUCTION

Until the beginning of the nineteenth century the Catastrophists envisaged Earth's history as a succession of abrupt upheavals, attributed to divine intervention and culminating in the Great Flood. In contrast, J. Hutton and C. Lyell favoured slow changes due to natural processes and interpreted Earth's history from contemporary evidence. Such Uniformitarians assumed that physical processes, both on the earth's surface and beneath it, are governed by invariable natural laws and are therefore essentially uniform and continuous. Uniformitarianism implies only that the basic laws are permanent. With different rates of working of the processes which have been understood, scales and rates of erosion and deposition change. Thus, ocean-floor sediments in the western North Atlantic represent the volume of 100 m of erosion in solid rock at the rate of 7 cm/1000 yr in the last 2.8 million years compared with 4 cm/1000 yr in the pre-glacial early Upper Tertiary (Laine, 1980), a significant contrast yet one entirely consistent with the main implication of Uniformitarianism.

During the course of the nineteenth century, Catastrophism gradually became the antonym for Uniformitarianism, supposing the operation of forces different in their nature from those observable at present, with disastrous ends to organisms and geological periods. Although these long words are much better left in the nineteenth century with other Victorian paraphernalia, the last decade has seen a sudden surge of interest in whether geomorphological and geological processes are, in a general sense, slow and steady or unexpectedly abrupt. Not least, the imagery of dynamic paleogeography of the 1970s, with the advent of tearing and colliding continents, sea-floor spreading, and commotion in the ocean, suggests that new wine is being funnelled into old bottles. Equally, the 1970s was the decade of the 'environmental impact statement' and witnessed the near drowning of Venice and half a rocky mountain blasting off. Clearly, the fundamental principle by which Earth's history is examined requires sustained, continuous assessment.

Emphasizing that physical, chemical, biological and astronomical phenomena act quite differently over spans of geological time, greatly clarifies the terminology of the abrupt, the gradual and the unexpected. Actualism is the implementation, when feasible, of A. Geikie's maxim that 'the present is the key to the past'. Gradualism describes the slow and steady operation of existing processes, exemplified by chemical weathering and continental drift. Postdiction or retrodiction is the antithesis of prediction and describes extrapolation from the present to the past. Paroxysms are violent or convulsive physical actions, with paroxysmal a very useful adjective since its literal meaning is 'rare'. People view paroxysms as natural hazards. A catastrophe afflicts living creatures and people. As the world becomes more widely and densely populated, the number of paroxysms that have catastrophic effect increases.

REGIONAL GEOGRAPHY

Striking clarification in contemplating the abrupt or gradual is achieved by careful attention to regional setting. The abrupt, in particular and unlike lightning, usually affects repeatedly definable and predictable locations and zones. Now that the margins of crustal plates are recognized as such, the strict geographical localization of abrupt subcrustal events affecting the earth's surface is clearly demonstrable (Fig. 2A–C). Such events may also show the equally geographical characteristic of change with distance (Fig. 2D), such as distance downwind from a volcanic vent (Fig. 2E) or distance away from an undercutting stream (Fig. 2F). In addition to the foci of vulcanity (Fig. 12A) the abrupt release of glacial meltwater is intrinsically circumscribed as an ice-margin phenomenon (Fig. 12B). In the striking case of Grimsvötn, geothermal melting of the ice adds to the individuality of the occurrence (Fig. 12C). Its discharge in the 1934 outburst was so great that for some hours it matched the flow of the Congo (Nye, 1976). The regional element in avalanches is equally crystal clear (Fig. 12D). Bushfires characterize seasonally dry, warm environments (Brown, 1972), and the wadi flood is the overquoted but rarely observed example of abrupt events in arid lands (Fig. 12E).

Geographical variability is also a characteristic of the ratio between the significance of gradual and abrupt events, since this varies considerably from one environment to another. Even adjacent areas may differ in responses to the same occurrence, depending on whether non-cohesive sediment is being transported or a solid landsurface eroded. For instance, sand is transported along beaches at relatively slow rates throughout a year, rather than by the occasional large storm waves which erode coastal cliffs. Similarly, in streams the largest portion of the total load is carried by flood flows that occur once or twice a year, but localized bank erosion and channel shift depend on rarer, larger floods. Paroxysmal stream-channel changes are a local phenomenon related to areally restricted cloudbursts, whereas a trunk stream tends to integrate such episodic events over a much larger drainage area (Baker, 1977). Elsewhere, braiding may be the result of flooding in historic times (Osterkamp, 1978).

In some areas extreme events may leave little mark on the landsurface. For instance, a flood following the June 1972 Hurricane Agnes in the Conestoga basin was four to fifteen times greater than previous annual flood maxima. Nonetheless, the ebbing flood revealed miles of virtually unaltered floodplain, only minor amounts of erosion, and virtually no mappable deposition (Baker,

---

Figure 12 Examples of abrupt natural flows

A   Contour map of deposits and location of glowing avalanche in the Quebrada El Pajal, following the October 1974 eruption of Fuego, a very active stratovolcano in Guatemala.

  (i) Contrast between irregular surface of glowing avalanche deposits in confined, upper portion of the valley, compared with channel-and-levée-like features where the valley opens out.

  (ii) Longitudinal profile of the path from notch in crater rim, descended at 50–60 km/hr.

(iii) Map showing track of descent path.

B   Observed drop in surface level of ice-dammed Tulsequah Lake on the eastern margin of the Juneau Ice Field in British Columbia. The fall occurred between 12.00 hrs on 6 July and 14.00 hrs on 11 July 1958.

C   Presumed route of ice-tunnel system beneath the Vatnajökull ice cap, draining from Grimsvötn, a subglacial, ice-dammed lake in south-east Iceland. About 75 per cent of the discharge is attributed to geothermal melting.

D   Velocities of the Slushman avalanche in the Bridger Range of southern Montana. The avalanche was triggered by the fall of a dry, hard slab of corniche, warmer, denser snow was picked up, and melting began in the run-out zone.

E   Flash-flood discharge in the Wadi el Washka, near Joufrah, 700 km south-east from Tripoli, Libya. The flood followed a 10 mm rainfall in November 1977, representing a quarter of the mean annual rainfall.

F   The Althea storm surge levels on the Queensland coast in December 1971, following a 10–18 knot cyclone. Damage in Townsville was estimated at 50 million Australian dollars.

Sources: Adapted from A Davies, Quearry and Bonis, 1978; B Reprinted – adapted – from Marcus (1960) *Geographical Review*, 50, with the permission of the American Geographical Society; C Nye, 1976, after Björnsson, 1975; D La Chapelle and Lang, 1980; E Lüder, 1980; F Hopley, 1974.

1977). Whether an abrupt event is a catastrophe depends on the localization of people and buildings, redoubling its geographical nature by depending on the areal association of nodes of human and of abrupt, natural activity.

ACCELERATION OF GEOLOGICAL TIME

To comprehend the temporal enormity of their subject, geologists telescope time at lightning speed. Although much can be explained by processes actually in operation, training and imagery conditions the mind's geological eye to see events at paroxysmal pitches. Poles wander, ocean floors spread, and the Pleistocene was a frosty blink. Compared with the 10 million years of Pliocence time, the huge Würm ice-sheet disappeared in a 10,000-year instant. In contrast, the Quaternary encompasses much of the time span which the geomorphologist envisages with some clarity, although an earlier interest in

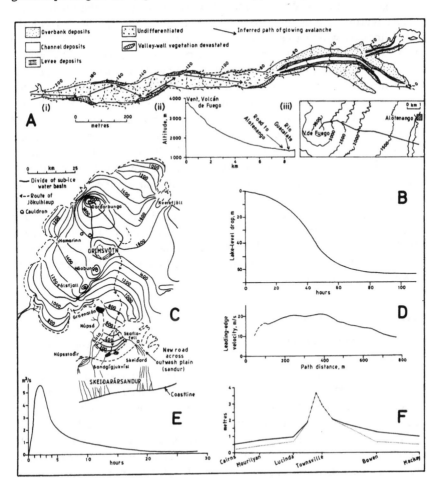

longer spans is beginning to grow again as continental drift lapses into the realms of established fact.

Some geological processes are rapid and abrupt, even within the context of a lifetime or a day's work. Anomalously rapid removals and accumulations suggest that occasional extreme events play a major role in the construction and modification of outer continental margins (Dingle, 1977). 'Active faults' are expected to be displaced within the span that buildings or engineering projects might remain serviceable. In particular, deep ice-core data are revealing occasional but very abrupt coolings within interglacials, one drop of some 5°C in fifty years probably being related to a cluster of volcanic eruptions (Flohn, 1979). In Greenland some 90,000 years ago there was an instantaneous change from a climate warmer than today to full glacial severity within a century, a plunge which took some 1000 years to reverse (Dansgaard *et al.*, 1972).

There are two geographical aspects to the acceleration of geological time. First, geographical scale is significant, since the largest of the earth's land-surface features are understood only in spans of geological time and with the use of imagery which accelerates time. Several such examples relate to 'failed arms', static rifts meeting at junctions where other rifts have moved or drifted apart. Thus, 'Some of the world's great rivers flow along failed arms, especially where nearing the ocean' (Burke and Dewey, 1973). Other drainages collected and focused by rift systems oblique to coasts include the Benue-Niger, Zambezi and Limpopo in Africa, and the Godavari, Mahanadi, Narmada and Ganges in India. In Europe, the Rhine flows along two arms of the 'Frankfurt junction' and in times of low sealevel reaches the ocean along the northern 'North Sea failed arm'. In the Americas there are the Amazon, Parana Mississippi, and the pull-apart rifting which controls the Rio Grande in Mexico (Potter, 1978).

The second geographical aspect is the marked regional variations in the spans of time perceived as geomorphologically relevant. C. D. Ollier stresses the Australian vision back to Precambrian times in contrast to much shorter spans in Britain. In the Netherlands, much before the Holocene or even historic time may be irrelevant.

HISTORICAL AND MORTAL TIME

Any description of the relative abruptness, intensity and recurrence interval of a geomorphological paroxysm must indicate broadly a span of time. Even a catastrophe is not necessarily moving more rapidly than people can run. The classic contemporary case is Venice, sinking at about 6 mm/yr and with the rate increasing; concurrently, the number of occasional high waters in the Adriatic ('*aqua alta*') is also increasing, with forty-eight of the fifty-eight recorded in the past century occurring in the 1935–70 period, and thirty in the 1960s (Berghinz, 1971). With some 70 per cent of Venice less than 1.25 m above mean sealevel, the implications are clear. If self-regulating mechanisms are absent in such cases, some slow movements may persist in one direction. These may outstrip technical advance and, with cumulative effect, may become a catastrophe. Thus, Earth's history, with sealevel rising 140 m between 15,000 and 5000 years BP, did terminate in a great 'flood' for unimaginative, uninventive and arkless islanders.

For extreme events occurring once or twice a year, recurrence intervals can be calculated accurately. Where the same extreme event has occurred twice, a recurrence interval might be estimated, assuming that such episodes themselves occur with a certain regularity in time. This has been observed in Norway for rockfalls on the Kiruna-Narvik railway between 1902 and 1960, is a basis tenet in the prediction of floods, but is less helpful for endogenetic phenomena. Occurrences within a lifetime are discerned in the context of normal time. Bearing natural hazards in mind, perhaps this temporal context is better described as mortal time. For historical time, however, implying a few centuries, some imagination is needed to speed up events.

PAROXYSMAL CHANGE IN MORTAL AND HISTORICAL TIME

There are five aspects to paroxysmal change which may be witnessed in a lifetime, recorded in history, or be deducible from the recent geological record. First, a potentially unstable situation develops, usually over a protracted period. Commonly, unconsolidated materials accumulate that would offer little resistance to downpour or flood. In Louisiana the 1957 storm surge elevated and transported entire portions of mudflats, including one segment 2 km long, part of the way across the adjoining marsh. Alternatively, where gravity is directly involved, protracted weathering finally brings a potentially unstable geological structure to the brink of paroxysmal disequilibrium. Snow accumulates to form avalanches in spring melts, or rivers and lakes may edge towards overspill levels. The overflow of Pleistocene Lake Bonneville flooded at about 280,000 m^3/sec, was over 120 m deep, and moved boulders more than 6 m in diameter. In some cases an initially small barrier, by intercepting more material, eventually releases a forceful surge when overwhelmed. This effect is created by log rafts, possibly an underestimated mechanism in many now-treeless areas. Over longer spans, active structural movements may lead to gravity-driven paroxysms. For example, in north-east India, the diversion of the Brahmaputra may have been partly due to gradual tilting of the Madhupur block.

Secondly, a triggering mechanism often finally releases gravity-driven movements. Above the surface, unusual atmospheric disturbances may be responsible. Earthquakes are a common trigger in tectonically unstable areas, as in the Mt Huascaran slide in the valley of the Rio Santa in north-central Peru on 31 May 1970. In very unstable features, triggering energy need only be small. In Hokkaido, falls of rain of only 50–100 mm are sufficient to trigger landslides. In many agricultural areas, human activity is more than sufficient.

Thirdly, a distinctive paroxysm is endogenetic, the volcanic eruption which combines the features of slow build-up towards the point of eruption with some trigger providing release. A spectacular example, the Minoan eruption of *c*.1470 BC took place on Santorini volcano. A caldera 11.5 x 8 km was created, and great volumes of ash and pumice were produced.

Fourthly, a natural chain reaction may be set up, often due to the constrictions of dissected mountainous relief where gravity-driven paroxysms are common. In 1958 30 million m³ of material fell into Lituya Bay, triggered by an earthquake along Alaska's Fairweather fault. This landslide then created a wave which swept up to an altitude of 530 m on the opposite shore. Earthquakes, landslides or volcanic eruptions may propel ocean waves hundreds of kilometres long at velocities of up to 800 km/hr.

Fifthly, the detailed geography of the location is a determining factor, as the Lituya Bay example demonstrates. Also, what is rare in one place may be an everyday occurrence in some instances. Mudflows are considered as normal denudational processes in the mountains of Japan because landsurface features resulting from these paroxysmal phenomena are observable everywhere in the landscape (Yoshikawa, 1974).

ACTUALISM AND POSTDICTION

With its abrupt and uniform changes compounded, the representativeness of the present for the past might be assessed. Strictly identical circumstances are, of course, not postulated. For instance, since most endogenetic processes are unobservable, the present is one number in a combination lock rather than a key. Within historical time, technological advances have become increasingly significant components of geomorphological processes. These include the widespread ploughing of recent decades and water abstraction from increasing depths, both of which complicate erosion-rate calculations. They range from contemporary building construction in suburban areas back centuries to the first forest clearances. On average, present erosion rates may be double those prior to the advent of technology (Douglas, 1967).

Post-glacial time poses several difficulties in making generalizations. Many episodes are punctuated by changes in the fossil pollen record which can be linked only notionally with counts of modern pollen rain. Crucially, it is hard to imagine that only 500 generations ago, sealevel was as much as 140 m below its present stand and that it then started to rise at a rate of about half a metre per generation. By the time that the rise ceased, more than one family may have counted livestock aboard two by two.

Throughout the Pleistocene the geologically unusual glacio-eustatic sealevel changes are not simply linked with climatic changes. Possibly the earth had never been so cold before. In the 20 million km² of formerly glaciated areas, the present is scarcely the key to the past in that the first ice-sheets crossed a ground-surface with a near-continuous mantle of weathered rock of very variable thickness (Feininger, 1971).

Further back in time, climate throughout the late Cenozoic was not typical of much of the geological past. Equatorial temperatures were somewhat similar, but polar regions, with the absence of land and consequently ice, were much milder. This caused much weaker meridional gradients than at present, perhaps resulting in a different and probably weaker general circulation of the atmosphere (Donn and Shaw, 1977). Even further, pre-Devonian relief was unvegetated and grasses appeared only in the Miocene.

The geological column itself, however, in general confirms the basis of geomorphological actualism. For instance, the composition of ancient greywackes coincides with the composition of contemporary river sands, suggesting broadly similar yields of debris (Potter, 1978). The validity of geomorphological postdiction is suggested by N. M. Strakhov who calculated that the average rate of sediment accumulation in basins of the geological past and the limits of variation of that rate fit within norms of present-day sedimentation. It is confirmed by the present disposition of highest-grade metamorphic rocks, crystallized at pressures equivalent to depths of 10, 20, or even 50 km. Their presence at the surface today implies colossal cumulative erosion subsequent to crystallization. Indeed, erosion has significant effects on the thermal structure of thick crust and may be an important factor in determining pressures and temperatures which are recorded by mineral assemblages.

Despite its many uncertain or incomplete aspects, postdiction is thus a significant if delicate art. Its dependability probably stems from the inexorable gradualism and scale of chemical weathering, which commonly accounts for over half of a denudational loss and also governs the rate of release of particles sufficiently loosened for physical transportation. Thus, a calculation of present-day erosion is multiplied by some time-span to indicate the interval over which some appreciable change in landform might take place. The order of denudation indicates whether forms are being produced at present or whether negligible rates suggest that they are relict features. Thus, a silica loss of 94 kg/ha/yr from a Dartmoor granite catchment (Ternan and Williams, 1979) is sufficient to imply that past processes of periglacial action or Tertiary deep-weathering need not be invoked to explain landform genesis of the controversial tors. In addition to arithmetic postdiction there is a second, distinctively geographical, aspect to postdiction. Knowledge from a study of currently active processes and associated landforms is transferred to areas where similar interrelationships may have existed in the past. Thus students of landforms in formerly glaciated areas invariably make first-hand observations on the margins of contemporary glaciers.

CONCLUSIONS

Modern plate-tectonic theory exemplifies actualism, being based on knowledge of the actual behaviour and present-day movement of existing plates. Precambrian wrench faults are compared to the San Andreas fault and ancient plate margins are deciphered a long way down the geological column. Thus, the theory is not just a paradigm for the pursuit of modern geology but also an emphatic re-establishment of its philosophical basis. For geomorphological processes an evaluation of the gradual with the abrupt may at first seem uncertain. However, the abrupt is usually infrequent in space as well as in time, and places where the abrupt is probable or even annual are readily located with geographical knowledge. Understandings of geomorphological processes are not necessarily directed to an interpretation of the past: particularly in unconsolidated materials, they are studied essentially as part of a dynamic present in which water or air rework materials. As a part of a human geography, the interrelationships between geomorphological processes and human activities is stressed. The sensationalism of the catastrophic lends morbid impetus to such anthropocentricism. However, if the future is to be predicted skilfully, there is no better training exercise than grappling with the uncertainties of postdiction. There are also natural processes with no reciprocal effects with people. Climates and sealevels gradually change, continents drift, volcanoes erupt or explode, and the earth trembles without any human assistance. Such processes, however, are equally important to the humanities because they prompt and challenge people to philosophize. In the Pacific where the Cococ, Nazca and Pacific plates meet, islands drift apart at 6 cm/yr and have rotated as rapidly as 10 degrees/million yr (Hey, Johnson and Lowrie, 1977). Today this is termed the 'Galapagos triple junction', the location where 'peculiar organic beings' were observed by Darwin in 1839. Today's ubiquitous question is whether mankind and our technology is a catastrophe, an '*aqua alta*' of increasing occurrence. As Mary Somerville (1848) observed, 'the change produced in the civilized world within a few years, by the application of the powers of nature to locomotion, is so astonishing, that it leads to a consideration of the influence of man on the material world' (Baker, 1947). Already four generations have been able to monitor this astonishing rate of change, to speculate about the implications of its intensity; and geomorphology continues to be an important source of accumulating data.

The Cycle of Erosion

THE CONCEPT AND ITS SHORTCOMINGS

> *The eye of an experienced topographer has discerned an accordance of summit levels signifying a former summit planation. (King, 1976)*

At the end of the nineteenth century, W. M. Davies proposed a Cycle of Erosion, adopting Lyell's view that paroxysmal convulsions are usually followed by long periods of tranquility. He postulated that initial, steep-sided incisions into an abruptly uplifted surface gradually broadened and flattened until an erosional plain of faint relief resulted, a hypothetical landsurface which he termed a 'peneplain'. Subsequently, the reality of peneplains have been much debated and the scheme was not universally adopted. Since no clear-cut, present-day examples are known, the argument for the peneplain was, quite simply, the antithesis of actualism. Even exponents found that Davisian cyclic interpretations within one area might be hard to match with those for other lithologies and structures, or from other areas.

A major weakness in implementing Davisian cyclic interpretations was the implicit use of an analogy between a postulated peneplain and the actuality of a stratigraphical horizon. Thus, wide, open valleys were hypothetically spanned between erosion-surface remnants just as gaps in a dissected horizontal stratum might be interpolated, a clear example of an unquestioned geological approach being inappropriate for landform study. These assumptions tended to ignore the gradualism of processes, by which erosion that might produce a later cycle would ensure that remnants of earlier stages would be progressively fewer and increasingly poorly preserved. Thus, even in the case of marine-cut notches, as in the case of Bonaire in the West Indies (Fig. 20C), only the lowest terrace is almost continuously preserved around the island. Gaps in terraces and variation in their width indicate that considerable erosion has taken place as each lower terrace in the sequence was formed (Bandoian and Murray, 1974).

Even if once present, the required morphological evidence may be undetectable to a non-committed eye. For example, off northern Scotland, few signs of erosion surfaces are visible on Shetland, and maxima on altitude-frequency histograms are largely due to variation in summit frequencies. 'If they are relics of erosion terraces, those terraces are old enough to have suffered very considerable erosion since they were formed' (Flinn, 1977). Ultimately, there is the great oddity in the argument that parallelism to the eye means peneplanation on the horizon, when the certainty is that the eye sees geometrical, parallel lines appearing to converge down the street.

THE ARID CYCLE OF EROSION

Arguments about the validity of the peneplain are largely insoluble simply because the Cycle and the peneplain were hypothetical constructs that only actually existed in Davis's imagination. This was not the case with the Arid Cycle of Erosion, which was modelled in part on the appearance of the Basin and Range province in the south-west United States. Here, recent seismic mapping and stratigraphic drilling have revealed the unusual distinctiveness of this area, emphasizing the inevitable obsolescence of Davisian generalizations based largely on morphological evidence.

The Basin and Range has long been recognized as a region of crustal

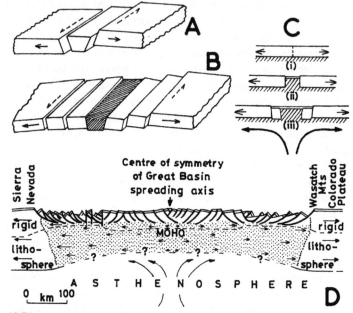

Figure 13 The explanation of rift systems in the context of plate-tectonic theory
A   The formation of a simple rift valley due to a relatively small amount of crustal extension. There may be considerable transcurrent movement.
B   The formation of the Red Sea structure, due to relatively large crustal extension.
C   Possible mechanism for creating the Red Sea structure and the separation of the crust. Rising mantle convection may induce tensional stress.
D   Approximate symmetry of Basin-and-Range faulting and possible diverging flowage at depth. Formation of the Basin-and-Range structure began 17 million–18 million years ago and extension has averaged nearly 1 cm/yr since, mainly along east-inclined faults in the west and on west-inclined faults in the east.
Sources: Adapted from A–C Girdler, 1965; D Proffett, 1977.

extension with normal faulting and differential vertical motion between the faulted blocks dominating the structure. Some basin floors, far from being erosional plains, have more than 3000 m of post-Laramide clastic sedimentary rocks, volcanic intrusives, and some anomalously thick bodies of evaporites proved. Geophysical studies reveal high surface heat flow and abnormally high temperatures at shallow depth. An extension of 30–35 per cent overall, or some 160–180 km across the entire Great Basin during late Cenozoic times is possible (Proffett, 1977). The axis of spreading in the central part is equivalent to spreading axes along which sea-floors are being created, and for which the Red Sea rift is the most studied example (Fig. 13A). The Basin and Range Province may therefore be a northward continuation, through the Gulf of California, of the East Pacific Rise (Menard, 1960). In the Dixie Valley area, spreading rates average 0.4 cm/yr over the past 12,000 years (Thompson and Burke, 1973).

Exterior drainage systems were developed, beginning some 6 million to 10.5 million years ago, and have evolved progressively to create the present-day landforms and drainage. This vivid regional detail emphasizes how little of later findings Davisian schemes could anticipate, and their inappropriateness as receptacles for present-day geomorphological knowledge.

PENEPLAINS AND UNCONFORMITIES

The Lake District mountains were it not that the destructive agency must abate
as the heights diminish would, in time to come, be levelled with the plains.
(William Wordsworth, 1822)

The marked breaks in the geological record represented by unconformities, and the peneplain hypothesis both raise several questions. A major concern is the relative rates of denudation and structural elevation. Equally important are the lengths of geological time when structural movements were small. Another question is the amplitude of residual relief that may persist on an erosionally 'levelled' surface. Finally, although demonstrable unconformities abound, even W. M. Davis admitted that it was difficult to point to a clear present-day example of a peneplain.

Using a denudation rate of 10 cm/1000 yr, Schumm (1963) calculated that it would take some 9 million years to lower the United States from its present mean level of some 690 m to zero. With uplift rates up to eight times faster than denudational losses, Schumm also indicated that ample time is available for erosion in the intervening periods of comparative stability. Further, many unconformities demonstrate an erosional reduction of rapidly elevated structures as a recurrent feature in geological time. Hercynian folding, for instance, was approximately levelled within some 20 million years of comparative stability. A relief amplitude of a few hundred metres on such structures is trivial compared with the geometrical reconstructions of the hypothetical structural relief. Paleorelief may be rugged and over 1000 m in scale, like the deeply eroded coastal batholiths in the foothills of the Western Cordillera in the central Andes (Noble, McKee and Megard, 1978). Nonetheless, the erosional reduction of orogenic structures to a paleorelief of 300 m, with local instances of 1000 or 1200 m, is an adequate datum for studying many stratigraphical problems. Thus, in western Canada, flat-lying basalts dated at 10 million to 13 million years were erupted on to 'a gently undulating erosion surface' that had a relief of 500–675 m (Rutland, 1973). Similarly, New Zealand's stratigraphic record suggests that peneplanation was widespread by Upper Cretaceous times.

Considerable variation in altitude is envisaged in the cyclic pedimentation interpretation of erosion surfaces of the African continent. L. C. King's highest 'pre-African' or 'Gondwana' surface is usually above 1200 m and is thought to be traceable into the Abyssinian Highlands at 2400–2700 m. It is ironic that W. M. Davis preferred to follow in Darwin's tracks to the Great Barrier Reef rather than visit the interior of Australia, one of the most stable continents, with no active volcanoes and large earthquakes rare, and with major diastrophism restricted to pre-Mesozoic times. It is a vast expanse of such low-lying land that the average elevation is less than 300 m (König and Talwani, 1977). Since it is merely a 30-million-year instant since Australia became detached from Antarctica, both its levelled surface and extensive duricrusts are more readily understood in terms of a Nile-like paleodrainage from headwaters in the Antarctica mountains.

Rough levelling down is undoubted for several large areas and at certain episodes in geological time. On the subcontinental scale, in central and western Australia, it is largely achieved. It is attempts to identify and correlate supposed flighted remnants of erosion surfaces within very narrow altitudinal ranges, and depending only on morphological evidence, which are now considered overoptimistic.

ALTERNATIVE INTERPRETATIONS OF 'CYCLIC' LANDSURFACES

Although applications may be localized, the array of alternative explanations, in non-cyclic terms, for landsurfaces bevelled across geological structures continues to increase. On the largest scale, geological structures themselves may give an impression of erosional truncation since some unconformities are now attributed partly to tectonic gravity gliding. More locally, but very commonly, resistant strata create local base-levels. Such instances have probably been underestimated and the number of regional base-levels overestimated, particularly in dissected plateaux. In the Appalachians, altitudes might owe more to two Carboniferous sandstones than to any succession of Cycles (Fenneman, 1936). More specifically, in the Little Swatara Creek, a much-studied catchment in Pennsylvania, resistant andesite lava underlain by shales at the outlet of the basin results in mean and summit levels being sharply distinguished 12 m above a lower shale basin (Meisler, 1962). At this scale and with proper attention to the heterogeneities of geology, few localities are without their own Little Swataras. Where a lake upstream is lowered progressively, the importance of local base-level and its stepwise lowering may be apparent. This is the case for strandlines in the Lake Victoria basin, associated with downcutting of the outlet, standing at 18 m, 12–13.5 m and 3–3.5 m, perhaps representing a 20,000-year fall.

For erosional landsurfaces not necessarily related to a local base-level there is a range of explanations for particular situations. For example, the *rampas* of south-east Brazil are slightly inclined planated surfaces which do not fit in with the Cycle of Erosion (Mousinho de Meis and Monteiro, 1979). Pediplanation is the explanation commonly adopted for erosionally bevelled landsurfaces in tropical areas, surrounding itself in its own system of controversy. Commonly, the emphasis in pediplanation is on the erosional retreat of the mountain or scarp front, rather than denudational lowering of the flatter landsurface areas. An example is the erosional bevels which occupy large areas of Guyana, which are similar to planations in Brazil and Venezuela and may be compared with features in equatorial and southern Africa. Such surfaces are regarded as pediplains cut back by denudation during periods of stillstand in the cyclic uplift of the Guiana Shield (McConnell, 1968).

Altiplanation, a concept formulated by H. M. Eakin in 1916 in Alaska, is sometimes invoked to explain level erosional steps in areas of active or former periglacial regime, with freeze-thaw dislocating the backwall and solifluction and meltwaters evacuating debris over the 'tread' of the step. Another alternative to peneplanation is suggested for bevelled escarpments and plateaux of limestones formerly overlain by an impermeable and largely insoluble cover (Pitty, 1968). Concentration of solutional activity at the surface could lead to differential lowering, the result of progressive exposure accompanying the gradual retreat of the edge of the impermeable cover (Fig. 14A). The application of this hypothesis to south-east England chalklands has been considered (Jones, 1981) and may be significant for interpreting the Mitchell Plain in southern Indiana where there is a gradual variation in bedrock exposed and the landsurface truncates the bedding structure with the St Louis Limestone outcrop (Fig. 14B). Generally, solutional loss is a significant proportion of denudation and an essential prerequisite for certain mechanical weathering processes. Therefore, ground-surfaces of any rock type more susceptible to weathering than an overlying retreating cover rock might similarly show the gradualism of differential lowering due to progressive exposure. This overall effect would, of course, be the result of greater dissection as well as greater downwearing, and a strictly non-planed surface is not envisaged in this model.

Compared with the hypothetical peneplain, features ascribed to being 'youthful' or 'rejuvenated' in terms of the Cycle are undoubtedly real. However, there are several ways in which local occurrences can be explained without cyclic suppositions. A particularly general feature is the effect of coastal recession which, by continuing to shorten stream length, may have the same effect as a fall in regional base-level. Thus, for streams draining to the North Yorkshire coast near Robin Hood's Bay, where coastal cliffs of Lias shales and sandstones continue to recede at 8.5–18.9 m/100 yr, continual steepening of stream profile is an adjustment to these changing conditions (Wheeler, 1979). This explanation is strikingly consistent with that now offered for a number of valleys dissecting the cliffs of the north Devon coast. Apart from the estuary of the Torridge and Taw, there are some fifty stream mouths. Only about a quarter of these valleys reach the coast at sealevel. All the other streams descend to the sea by waterfalls, varying in height from a few metres up to the three-storey fall of Speke's Mill Mouth, with its total drop of some 45 m. Such features can be explained by cliff erosion dismembering former valley systems running parallel to the coastline (Dalzell, 1980). Comparable low-lying 'marine platforms' are currently, on a geological time

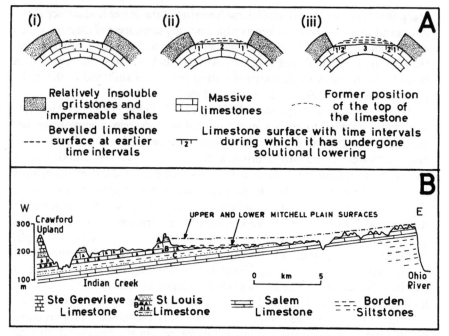

Figure 14 Illustrations of how progressive exposure, leading to differential solutional lowering, could produce a landsurface bevelled across a limestone outcrop
A   Model based on chemical weathering, soil residue thickness, and morphological data from the southern Pennines.
B   Upland surfaces in southern Indiana which truncate the bedding structure of limestones undergoing exhumation, an instance where differential solutional lowering due to progressive exposure could be considered.
Sources: Adapted from A Pitty, 1968; B Palmer and Palmer, 1975.

scale, being produced in a similar way at the seaward extremities of the Dingle Peninsula in south-west Ireland.

Looking inland, with a geological time-telescope trained on inclined sedimentary strata, one sees escarpments being rolled back like carpets. In detail, drainage from a higher basin may be diverted into a lower system without any 'capture' mechanism linked with regional base-level, but simply by the lateral erosion of one side of a valley encroaching on the heads of tributaries in an adjacent system. This is postulated to be the case of downdip shift and breaching in inclined strata, particularly feasible where the retreating escarpment has a dipslope already dissected by dip-parallel drainage (Pitty, 1965).

CONCLUSIONS

In recent decades two directions of research have moved investigations away from the successes and limitations of the Cycle of Erosion. First, the advent of techniques of absolute age determination have removed the main necessity for attempts to establish the age of landsurfaces from their appearance and altitude alone. Such techniques, and much subsequently acquired and more precise evidence, relieve the geomorphologist of the need to question and debate whether the generalizations in the Cycle of Erosion concept are sufficiently precise to support quantitative descriptions. Instead, geomorphological measurements and observations can be made with freedom from the foregone conclusion that these should fit into a cyclic frame.

Secondly, the expansion of interest in aspects overlooked in the Cycle provide some of the more obvious examples of interests with increasingly large followings. These include process studies in general, but with particular value placed on research into processes in environments other than that where the 'normal' Cycle was thought to apply: Quaternary geology, slope-form studies, hydraulic geometry, and soil and sediment studies. Also, contemporary laboratory facilities offer opportunities for work with sophisticated instruments which invite research away from topics, such as the Cycle of Erosion or the features it sought to explain, to which their application is not readily apparent.

Despite the attractiveness and success of post-Davisian geomorphology, two issues surrounding the Cycle concept remain essentially unanswered. First, the length of time over which the Cycle held sway may seem to reflect some fundamental validity. The second issue is that no concept of comparable generality has been amicably agreed. Four possible reasons for the extended span of the Cycle's popularity can be offered. First, Davis's admirers were perhaps too uncritical of their patron's work, despite the strong challenge to the Cycle from leaders of German geography like Hettner and Passarge (Tilley, 1968). Secondly, some of the premises of the Cycle run counter to basic scientific tenets of geology. With no present-day examples to cite, actualism is ignored in the peneplain concept. Equally, in assuming that remnants of former peneplains might be identified and linked as reconstructed surfaces at several altitudes, the gradualism of continuing denudational processes had to be overlooked. It is difficult to counter in logical terms such a scheme in which basic principles are simply ignored. A third reason was possibly the compelling, directly personal appeal of the terms Youth, Maturity and Old Age. As inescapable facts of life, the possible irrelevance of these phases to the inanimate world is not immediately obvious. Finally, the Cycle has also persisted because, in a less idealized way than Davis's formulation envisaged, denudational lowering of geological structures is inevitable and proceeds inexorably.

The fact that no concept of comparable generality to the Cycle has emerged is probably due mainly to Davis overtly rejecting the basic geographical tenet of regional variability, as apparent in the phrase 'Normal Cycle of Erosion'. Quite simply, much of the subject matter of landform study is too variable geographically to be accommodated within such wide-ranging generalizations.

Climatic geomorphology

INTRODUCTION

The Normal Cycle soon encompassed an arid and then a glacial cycle too. A generation later, nonetheless, Cotton (1941; 1942) still represented departure from Normal erosion as 'climatic accidents'. At mid-century, two new schools of thought were formalized, both emphasizing process, each anxious to provide an alternative to the Cycle of Erosion but, in terms of their perceptions of the significance of regional climate on landforms, proceeding in opposite directions. In some cases it was noted that the physical principles involved in weathering and in resistance to weathering, although they may vary in intensity from region to region, are immutable. A basic statement was by Horton (1945), an engineer who noted that

> The geomorphic processes we observe are, after all, basically the various forms of shear, or failure of materials which may be classified as fluid, plastic or elastic substances, responding to stresses which are mostly gravitational but may also be molecular . . . . The type of failure . . . determines the geomorphic process and form.

As physical laws are immutable, it is deduced that landforms will tend to be similar regardless of latitude. This physical approach is that of some leading American geologists, such as Strahler, who makes relatively few references to climate, and Leopold, Wolman and Miller (1964), who conclude that 'evidence indicates that the forms of hillslopes may be identical in all climatic regions'. This echoes the observation that 'all forms of hillslope occur in all geographic and climatic environments' (King, 1957).

In contrast to the physical-principles approach, the alternative conclusion was reached by an environmental or climatic approach to geomorphology. Here the interest in process recognizes differences in physical factors and extends to include the unobservable chemical and biological processes. The hypothesis is that the well-established zonal contrasts in climate, with their distinctive effect on vegetation, mean that physio-chemical processes combine in varying ways and operate at different rates, with interrelated vegetational influences accentuating the distinctiveness of morphogenetic processes in contrasted regions. The immutability of physical laws is less important than the environmental contrasts for landform development, and it is assumed that the latter should leave some discernible traces on the landsurface form. The emphasis is therefore on studying process and landform in various environments, with the accent on establishing the changes in the relative effectiveness of biochemical to physical weathering.

A minimization of the importance of latitudinal contrasts in climate is perhaps an inevitable conclusion which follows from the initial assumption that physical considerations provide the basis for landform and process study.

The neglect of biological and chemical considerations means that the very factors which do change with latitude are omitted. The most clearly defined landforms are due either to deposition or to biochemical recombination of weathered elements such as glacial deposits in higher latitudes. In lower latitudes there are coral islands and the plateaux effects induced by duricrusts.

MORPHOGENETIC ZONES

More specific than an interest in climate on landform is the genetic emphasis and the regional character, implicit in the word 'zone', of the concept of morphoclimatic zones. The distinctiveness of climatic regions and its possible implications for landform study were first considered by A. Penck in 1883, and he attempted to classify correlations between climatic areas and surface relief in 1910. In that year E. de Martonne was probably the first to introduce the term 'climatic geomorphology'. Its later progress owes much to German workers after Penck, like C. Troll, and J. Büdel. The tacit assumption is that from the distinctiveness of the environment it follows that the forms must be distinctive. In this way some morphoclimatic types have been defined on climatic data alone without initial reference to the forms which are then fitted, or assumed to fit, into the scheme. Other schema avoid Büdel's actual delimitation of regional boundaries of five morphogenetic zones and merely produce a physical geography catalogue of climate, vegetation, soils and landforms for a recognized climatic zone, with little attempt to link these into a morphodynamic system. The work of A. Cailleux and J. Tricart, searching for the essential characteristics of an area rather than speculating about hypothetical, indefinable, if not bizarre, boundaries, is very much in the tradition of French regional geography. Their attempt is to recognize distinctive morphogenetic processes and to link these to landforms of a given zone, often with sediment morphometry as an important aid.

Apart from the physical school, who essentially ignore climatic geomorphology, several workers are unconvinced. Derbyshire (1973) reflects that the demonstration of the concept by Büdel is imperfect. Indeed, as with the Cycle of Erosion, setting up world-wide schemes in advance of sound factual bases and then searching for the factual confirmation is an unsound procedure (Stoddart, 1969). One particular problem is too little regard for structural control (Kennedy, 1976).

The overoptimism with which climatic geomorphology was launched can be linked with the disciplinary problem of geography that its successful application would solve. Its conception reflects the belief in the reality of identifiable regions where this still prevailed in the post-1945 years. Equally, the danger of overemphasizing the unique is an inherent danger in climatic geomorphology (John and Sugden, 1975). For example, studies in the periglacial zone are very much concerned with processes and morphology unique to high latitudes, whereas azonal fluvial action of closely spaced washslope rills transports significant quantities of silts after snow melt (Wilkinson and Bunting, 1975). As a possible alternative to the Davisian approach, it was too readily embraced as apparently a different means of explanatory landform study. Later, American geographers briefly hoped that Büdel's approach would reveal a real variation and spatial pattern and thus merge conveniently with their changing emphases (Holzner and Weaver, 1965).

SOME EXAMPLES OF LATITUDINAL VARIATIONS

It is important to give some examples of geomorphological characteristics which are linked with climate before examining further the particular difficulties of generalization. Figure 15A shows a range of drainage densities for areas ranked according to Büdel's morphogenetic zones. It shows that both the density values and their range increase from mid-latitudes, through sub-tropical to semi-arid zones. These decrease further in the arid zone before increasing once more in the tropics. In Figure 15B Toy (1977) presents the generalized change observed in several measured slope profiles on marine shales, half lying close to the 37th parallel and the remainder extended along the 105th meridian, showing a clear difference in erosional environments. The slope length and shape was intimately associated with stream density which was greater in the semi-arid than in the humid areas. This striking demonstration may not, however, be detectable on more resistant strata. Figure 15C illustrates the inevitable tendency for permafrost and glaciation phenomena to occur at lower altitudes in higher latitudes. It also shows a parallelism between permafrost limits and the tree line. Although the treeline appears to be a basic discontinuity in the geomorphological environment of a mountain region, there is no support for a morphoclimatic boundary at the timberline in mid-latitudes (Caine, 1978). Figure 15D shows the significance of continental

---

Figure 15 Examples of geographical contrasts and trends in climate, associated with changes in process, forms and sediments
A   Selected values and ranges of drainage density in Büdel's contrasting morphological zones, ranked latitudinally.
B   Generalized contrast of hillslope shapes in marine shales between arid and humid regional climates. Field measurements were made at twenty-nine sites in the United States along two traverses, half close to the 37th parallel, the remainder on the 105th meridian.
C   Morphoclimatic zonation in the Rocky Mountains, suggesting parallelism of permafrost limit with tree line.
D   Approximate latitudinal zonation of inner-continental shelf-bottom sediments. Such data summarizes the effects of denudational regimes on adjacent land, such as mud being most abundant off zones of high temperature and high rainfall in the humid tropics, sand increasing in the hot, arid subtropics, and rock and gravel predominating off the coldest zone.
Sources: Adapted from A Gregory and Gardiner, 1975; B Toy, 1977; C Harris, 1979; D Hayes, 1967.

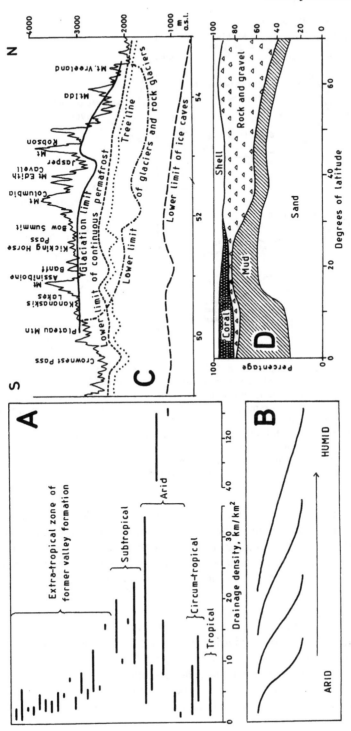

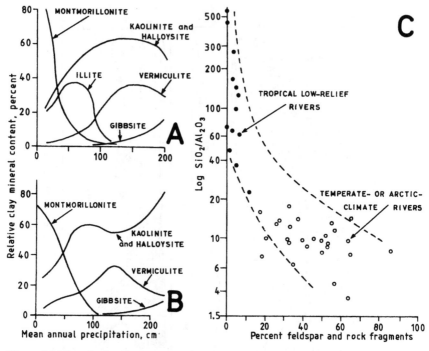

Figure 16 Climatic significance of clay and sand mineralogy
A and B   Clay mineral formation on different rocks in California, in relation to a range of mean annual precipitation. Only acid igneous rocks give rise to illite at any rainfall (A); Montmorillonite remains stable at higher rainfalls where the parent materials are basic igneous rocks (B).
C   River sands with high quartz percentages and high SiO_2/Al_2O_3 ratios are chiefly from low-relief tropical river basins (closed circles). Open circles represent river basins with either high or low relief in either temperate or arctic climates.
Sources: Adapted from A and B Singer, 1980; C Potter, 1978; 444.

shelf sediments for reflecting the balance of the denudational regime on land, with mud predominant in the high-temperature and high-rainfall zone of the humid tropics. This compares with the rock and gravel of cold environments and with the volume of sand being greatest in the hot, arid subtropics. Certain changes in sediment type can also be observed on land. Figure 16A shows the inverse relationships between clay types in Californian soils depending on amounts of precipitation; it also reminds one of the importance of lithology, since illite occurs only on acid igneous rocks. Conversely, Figure 16B, in showing the climatic significance of rivers and petrology and chemistry, also demonstrates that high quartz and high SiO_2/Al_2O_3 ratios are chiefly from low-relief tropical river basins where weathering is sufficiently intense to override any differences in source rocks. Thus, several particular phenomena can be seen to differ with climatic zone. The questionable assumption is whether such changes might be areally associated with others and if their grouped characteristics could form a mappable unit.

It seems that climatic geomorphology is another example of the concepts of low explanatory power increasingly recognized in this discussion. Four aspects of morphogenesis which contribute to this low explanatory power are the complexity of climatic influences, paleoclimates and paleoforms, multizonal processes and forms, and convergences.

COMPLEXITY OF CLIMATIC INFLUENCES

The basis of the morphoclimatic proposition is the equatorward increase in average temperatures, reinforced by higher potential evapotranspiration. However, receipt of solar radiation is less clearly zoned, and can be intense from cloudless skies in deserts or high mountains, or during long arctic summer days. More obviously, the irregularity of precipitation belts complicates latitudinal progressions in weathering and transportation processes.

Altitude greatly complicates the morphoclimatic character of a given latitude. Thus, in central Africa, Kilimanjaro (5879 m in height), Mt Ruwenzori and Mt Kenya all carry glaciers. Cirques with floors at about 4750 m surround the most easterly of Kilimanjaro's volcanoes. The snow line is about 4500 m on Ruwenzori which compares with about 5000 m in the equatorial Andes. The disposition of mountains on the western side of continents and the direction of prevailing winds create arid patches rather than belts. In areas like Indochina, southern China and Atlantic Brazil, forest cover continues from equatorial areas to humid mid-latitudes, whereas in Africa a broad arid area intervenes north of the equator.

Oceans may be a morphoclimatic embarrassment, particularly on J. Büdel's maps. In higher latitudes, periglacial activity is practically without zonality, partly because in the southern hemisphere there is virtually no land in the appropriate latitudes. In the northern hemisphere there are contrasts like those between southward-jutting Greenland and the coasts of north-west Europe influenced by the north-east extension of the Gulf Stream.

Large rivers may introduce substantial changes in flowing through areas which contrast with their headwaters. In the northern hemisphere there are contrasts between north-flowing and south-flowing major rivers. In tropical areas several smaller rivers create conditions like the Nile on a local scale, with heavy rains in catchment headwaters maintaining humid zones through arid lowlands. For example, in equatorial Somalia, the vegetation is generally dry savanna, through which the river valley is a 2–3 km-wide densely forested strip resembling the humid tropical environment.

PALEOCLIMATES AND PALEOFORMS

The concept of morphoclimatic zones tacitly implies that landforms are largely the result of present-day processes, yet climatic oscillations are the overriding characteristic of the Quaternary. In central North America, isotherms shifted south by at least 1200 km in the ice-covered areas. Glaciers were once present in tropical highlands like the East African Mt Elgon, Aberdare Mountains and the Abyssinian Highlands. In tropical lowlands like West Africa abrupt climatic changes produced coarse gravel terraces, which resembled the results of periglacial episodes of the humid-temperate and Mediterranean zones. Further back in time, climatic norms drifted as the distribution of continents and oceans changed significantly during the Cenozoic. Even zonality itself was less pronounced. There were pines and firs in Greenland in the Oligocene, paleotemperatures of 15–25° have been surmised for the Antarctic Peninsula in the late Eocene, and there is evidence of Tertiary tropical flora on the Canadian Shield.

Since morphogenesis may be quite slow, particularly for larger-scale forms in resistant rocks, paleoforms may persist as relicts from times when a different set of processes prevailed. For example, the classic pediments in the crystalline rocks of the south-west United States deserts have been attributed to arid-zone processes which are not currently operative in granite terrains in the Mojave Desert (Oberlander, 1972). G. H. Dury is confident that the interpretation of paleoforms in terms of a past climate is the main contribution of climatic geomorphology. The paleoforms of periglacial activity attract much interest. For instance, frost creep and solifluction produce distinctive lobate and terrace-like landforms. Recognition of 'fossil' deposits must be based upon a combination of characters including microform expression, sorting, structure, fabric and associated thin, buried organic horizons (Benedict, 1976). It becomes difficult to distinguish forms of different origins after modification by runoff. A typical example of these valued indicators of the former extent of periglacial activity is the interpretation of earth mounds as relict pingos in many parts of western Europe, as near Llangurig in central Wales. A 'pingo', an Eskimo term for ice-cored earth mounds, may collapse in the centre, with the remaining segments of the rim holding up a shallow lake. The development of the mound is attributed to interpermafrost and subpermafrost water seepages coming to the surface under hydrostatic pressure (Fig. 17A). Water temperature must be very close to 0°C and if the volume is too large it will remain unfrozen as a spring. If it drops below freezing, the seepage is sealed off (Holmes, Hopkins and Foster, 1968). Examples interpreted as relict pingos by Watson (1971) occur in the Cledlyn basin in central Cardiganshire. Their location is similar to those represented in Figure 17A. The outer face of the ramparts rises 5–6.5 m above their base. Generally overlapping, they include more openly spaced, elongate features at their western end (Fig. 17C). In contrast, shallow and often overlapping depressions in western parts of East Anglia on the Chalk between boulder clay and the Fens do not resemble the much larger and sparsely distributed pingos of contemporary arctic environments (Sparks, Williams and Bell, 1972). Comparison with the central Wales examples is difficult due to the low relief and changed groundwater conditions

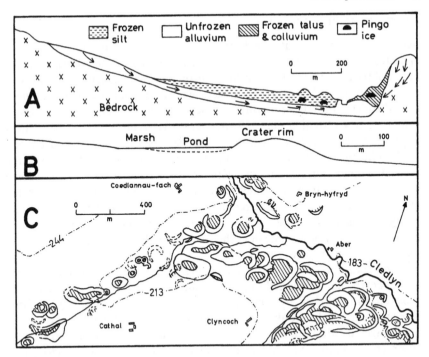

Figure 17 Example of a comparison of contemporary, developing landforms with postulated relict counterparts – pingos in Alaska and comparable forms in central Wales
A   Typical location of pingos in central Alaska. The arrows represent groundwater movement; vertical exaggeration 2×.
B   Profile of a typical pingo in Alaska, in an advanced stage of collapse. Vertical and horizontal scales equal.
C Semi-circular ramparts and impounded marshes, interpreted as relict Pleistocene pingos, on the Cledlyn valley floor, central Cardiganshire.
Sources: Adapted from A and B Holmes, Hopkins and Foster, 1968; C Watson, 1971.

typical of Chalk. However, the possibilities of relict pingos, despite their small and scattered nature drawing apparently undue attention to the unusual, exemplifies the scope and uncertainties of climatic geomorphology viewed as the interpretation of relict forms.

Although climate may leave only occasionally identifiable scratches and smears on the landsurface rather than a bold imprint, and despite substantial and repeated oscillations in climate, the concept that some areas may have remained essentially morphoclimatically stable over extended periods of time may be worth retaining for certain areas. Thus, Mensching (1970) contrasts the Central Sahara with the 'unstable' peripheral Sahelian and Sudanese zones. Similarly, M. F. Thomas (1974; 285) regards it as entirely logical to regard certain landscapes as the product of a limited range of environmental conditions.

MULTIZONAL PROCESSES AND FORMS

Several processes are multizonal. Running water is a fundamental process in all climates beyond permanently frozen areas. Even relict drainage patterns like North African wadis reflect this fact. Distinctive regimes can be azonal, with intermittent flow characterizing limestone terrains, semi-arid areas, and periglacial zones. Permafrost produces the same surface effect as hardened clay or solid rock in arid-region slopes. As neither surface is protected by a permanent vegetation cover, sporadic or seasonal supplies of surface water produce sheet-wash. The heavy sediment load associated with great fluctuations in discharge favours aggradation and the development of braided stream courses. Some vegetation characteristics which can influence landforming processes are multizonal. For instance, the predominance of roots in the tundra community (70–80 per cent of the biomass) is like arid communities. On both environments, conditions are so harsh at the surface that living matter is best able to maintain activity below ground.

Wind transportation is active throughout polar deserts, particularly where no pavement has developed to protect sandy material. Bedrock may become grooved and polished by wind action. Even chemical weathering can have some uniformity. In the absence of water, salt encrustations and lakes occur in north Greenland, as in hot arid deserts, due to strong evaporation in mid-summer. Similarly, in McMurdo dry valleys in Antarctica, salt weathering prevails, with cavernous undercutting causing free-face recession (Selby, 1971).

'CONVERGENCE' OF FORMS

As well as process effects, many distinctive features and forms are reported equally from a wide span of latitude. Such forms may be the result of the same azonal processes or, alternatively, two contrasted sets of distinctive processes can product the same form. Equally, paleoforms may persist, relicts from when a given set of processes different from that of the present day once prevailed. It is therefore circumspect to emphasize the possibility of different processes producing essentially similar forms. King (1953) wrote of homologies in landform, Wilhelmy (1958) of convergence. The latter word appears often in the writings of J. Tricart and A. Cailleux. The American geologists term 'equifinality' is similar but does not carry the implication that the convergence is between forms from different climatic weathering regimes.

Some examples of homologies include microforms produced by rock weathering in all types of climate, as described by Wilhelmy. They are commonly associated with the various mechanisms which combine to produce changes in volume. Cracks are functional in the development of some soil patterns in both cold and warm climates, and expansion leads to bumps as widely separated as the Icelandic thufurs and the Australian gilgai mounds. There is a close similarity between polygons produced by desiccation in clay and those in frozen ground. On a larger scale there are the semi-arid and periglacial ramp-like lower hillslopes, and Birot (1966), in describing the dissection of the Serra do Mar escarpment in Brazil, notes that the valleys have some similarity in appearance to the U-shaped glacial valley. More recently, tower forms in limestone up to 125 m in height, previously thought to be distinctively humid, tropical karst forms, have recently been discovered in the Mackenzie Mountains of Canada's remote North-West Territories (Brook and Ford, 1978).

The rather important-sounding words 'equifinality', 'convergence' and 'homology' must be used carefully as their connotation varies with viewpoint. If equifinality is a recognition that the laws of physics are insufficient to predict the shape of the landsurface, equally convergence and homologies may imply that the general is unexpected.

CRITICAL EVIDENCE FROM COASTS

In attempts to elucidate the climatic geomorphology hypothesis, its several complicating factors have traditionally been reduced in an important direction by considering the case of karst forms (Sweeting, 1972). This eliminates some of the variability attributable to lithology and simultaneously focuses attention on one process, that of solution. The implications of Brook's and Ford's discovery may take some time to evaluate. In the meantime, the clarification of the nature of the climatic geomorphology concept which is obtainable from coastal studies, can be reviewed. With constant supply of water and energy, the physics of beach and shore is uniform in general principle and in the dynamics of operation. Swales, cusps, spits, tombolos and other features are found in many latitudes. However, beaches in cold climates have additional features. Pitted beaches occur where buried ice melts, frost cracks some hundreds of metres long are common, and stone circles and polygons may result. Ice-pushed ridges may occur. Not least, preservation of abandoned ridge and swale systems over decades is unique to the Arctic (Short, 1975). Mechanical reworking by waves may be restricted for nine to ten months of the year when coastal fringes remain frozen by spray from surf, and swash freezes to form an ice crust on the shore and lower parts of cliffs. Disturbance by severe storms is rare. The coarse fraction forms a gravel pavement, preventing further deflation. Permafrost maintains perched lakes in the swales and the water also prevents deflation. In lower latitudes, eolian and storm activity tend to rework berms and beach ridges continually.

Unconsolidated and recombined materials vary with latitude. Frost weathering or glaciation leave large quantities of coarse material on adjacent shores. Solifluction lobes may encroach on beaches. Boulder-strewn coastal flats are restricted to cool temperate regions, like the St Lawrence estuary, where drift ice is important in transportation and deposition. There is little inorganic material on the shores of hot, arid lands where runoff is limited and allogenic streams absent. Coral sands, coral growth, beachrock and other biochemical phenomena are restricted to lower latitudes. Intertropical coasts are fringed with huge cordons of sand, with swamps inland, due to the abundance of fine sediment in rivers. Headlands are infrequent, and coarse particles rare, due to deep weathering of many rocks. Mangroves flourish in regions of tropical climate where rapid accumulation of fine sediment occurs. The position of the shoreline shifts continually. In Indochina, Venezuela, Guinea and Brazil the coasts reveal periods of rapid accretion followed by shorter intervals of wave and current action (Vann, 1980). Mangroves may line some 20 per cent of the world's shores, particularly along the north coast of South America.

CONCLUSIONS

A minimization of the importance of latitudinal contrasts in climatic morphogenesis is perhaps an inevitable consequence from the initial assumption that physical principles provide the framework for landform and process study. If the equal significance of biological and chemical phenomena is overlooked, the very features that do change with latitude are omitted. On the other hand, early notions of homogeneous belts of distinctive processes and associated forms were quite misleading.

Today, the precedent established on coasts is for distinctively but usually minor touches to be recognizable in depositional forms. Comparable precise contrasts have yet to be established for erosional landforms inland. If present, they await deciphering by detailed measurements, such as those of T. J. Toy (Fig. 16B). As intermediaries between form and process, perhaps clay mineralogy of soils deserves closer if not prior attention. Clays influence hydrological and mechanical properties of the ground-surface and exhibit the broad but significant contrasts observed in zonal soils on continental plainlands for up to a century. In Africa, as on most ancient, much-weathered landsurfaces, the soils are essentially kaolinitic, whereas montmorillonite is typical of the clay fraction of warm, arid zones (Pitty, 1979).

If climatic geomorphology has failed to please those geographers requiring areal variation and spatial pattern, its 'convergences' and relict forms appeal strongly to those following the traditional comparative method and exploration, as exemplified by Brook and Ford (1978). Overstated, climatic geomorphology 'can be viewed as an unfortunate diversion of attention from the far more significant landform/structure relationship' (Clayton, 1971). At best critically useful in certain circumstances, it reminds us again that geomorphology is a subject in which explanation is of such sufficiently low power to tempt overemphasis from enthusiasts of a concept and its dismissal, without great handicap, by others. Thus, a realistic summary of contemporary morphoclimatic approaches is almost synonymous with 'process-orientated geomorphology', but specifies studies within an environmental context of distinctive climate, soil and vegetation cover, past or present.

Stillstands and the mobility of earth structures

Between the crust of the earth and the forces of erosion there is an unequal
interplay that has brought into being landforms of extremist variety.
(Bowman, 1926)

INTRODUCTION

The Davisian Cycle depended on assumed protracted periods of stillstand of
earth structures to run its course. Conversely, mobility of the crust was a
principle on which W. Penck founded his approach. The major, immediately
recognizable, broad zones of the earth's surface are those of structural mobility
which tend to put the small-scale refinements of climatic geomorphology into
perspective. Regardless of either cyclic or morphoclimatic considerations, the
recent advances in knowledge of the stability and mobility of earth structures is
vital to several aspects of landform study, with its intrinsic regionalization a
striking feature. The topic is of interest to geologists concerned with transgressions and regressions. Equally interested are man-orientated geographers
since a few metres up or down in the next few decades or centuries has
far-reaching implications. Interest is redoubled by the possibility that man's
current effects on climate are capable of inducing such changes with rising CO_2
levels in the atmosphere providing the so-called 'greenhouse effect'.

Davisian geomorphologists were much concerned with a base-level below
which most mechanical erosion ceases. Sealevel provides a general base-level,
perhaps notched at the coast by marine erosion. Inland, base-level is a hypothetical datum towards which the downward degradation of peneplanation has
progressed. However, in most areas, and particularly in uplands, a resistant
rock outcrop commonly sets a local base-level for landsurfaces upstream.
Much of the recent research on processes is localized within such areas, so that
with the effects of the demise of Davisian geomorphology and the rise of
process-orientated work, interest in base-levels has experienced a double
decline.

Three aspects of the continuing study of stillstands and mobility of earth
structures are relevant to geomorphological enquiry. First, base-level is
important, even in non-cyclic terms and with local significance only, for the
evolutionary history of the lower reaches of any stream reaching the coast, its
valley-side slopes, and any terrace remnants that persist. More broadly, any
relative uplift of land or equivalent fall in sealevel will inevitably intensify
erosional activity, especially if base-level change is rapid. It is interesting to
consider whether there are significant lags in increased denudation after uplift,
and whether the greater relief is eventually a source of sufficient energy for
denudation to match uplift. This possibility is readily examined in areas of
isostatic rebound. Thus, where the Sault Ste Marie rock barrier emerged to
separate Lake Superior from Lake Huron about 250 BC, downcutting

approximately matched this rate (Fig. 18A). A longer-term situation is present in the Japanese Mountains which have been uplifted at a near-constant rate since the beginning of the Quaternary. Figure 18B shows that the greater the uplift, the more intense the erosion. However, due to the other factors such as rainfall intensity and vegetation cover, the denudation rates corresponding to a given amount of Quaternary uplift range fairly widely. Also, there is consider-able regional variation in the amount of uplift and a tendency for denudation to exceed contemporary uplift rate in certain areas, such as the Central Mountains and the Outer Zone of south-west Japan.

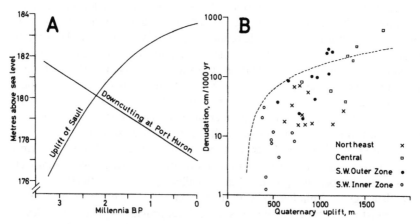

Figure 18 Examples of the simultaneous occurrence of uplift and denudation
A   Downcutting and isostatic rebound at the Sault rock barrier in St Marie's River, Ontario. The rock barrier emerged after the Algorna water-level stage, *c.* 3200 BP.
B   Quaternary uplift and contemporary denudation in Japan. On the dashed line, contemporary uplift and denudation are equal.
Sources: Adapted from A Farrand, 1962; B Yoshikawa, 1974.

Secondly, many difficulties in interpreting and interpolating erosion surface remnants are now being rediscovered in a narrower context. In structurally mobile belts, coastal slopes occasionally incorporate flights of narrow erosional notches, similar to wave-cut platforms, at their base. These relict platforms are one of many lines of evidence that help to elucidate past relative levels of sea and land in such areas. Other features include uplifted stacks and knickpoints in stream profiles, as observed in Japan, New Zealand, New Guinea, western North and South America (Morisawa, 1975).

Thirdly, the broader purpose of such studies is decidedly geological, aimed at establishing the degree to which marine transgressions and regressions over continental margins were synchronous over widely separated areas. The global nature of such advances and retreats of seas, such as occurred in the Upper Cretaceous, has been recognized for nearly a century. Such studies, however, have offered geomorphological glimpses of the earliest landsurfaces to have some bearing on the evolution of present-day drainage patterns.

TECTONIC FRAMEWORK AND ACTIVE AREAS

The degree to which portions of the earth's crust subside, rise or are stable has three aspects. First, compression can result in substantial vertical displacements in strata susceptible to broad-scale folding. Elsewhere, epeirogenic movements are primarily vertical and affect continental blocks and also stable zones of former folding. Finally, cratons are stable portions of continents that have resisted deformation for prolonged periods of geological time. Broad contrasts are discernible on a continental scale. For instance, the main uplift occurred as recently as late Miocene in the Andes. On average, Africa is 50–200 m above the other continents due to its epeirogenic uplift compared with subsidence of other continents. Several crustal swells or domes of Tertiary age rise conspicuously above the surrounding relief (Bond, 1978).

For both folded zones and platforms, vertical movements are of the order of tens to hundreds of centimetres in a thousand years. In the Los Angeles basin, terraces are currently rising at some 400 cm/1000 yr. Africa, which incorporates several cratons, experiences vertical movements of anything between nil and 420 cm/1000 yr. Extreme values in the most active tectonic regions of the world include elevations of 1630 cm/1000 yr along parts of the Caspian Sea. In fact, the areas where current movements are part of huge disturbances starting in late-Tertiary and within Quaternary times are regionally circumscribed. For instance, great thicknesses of clastic fill suggest downward displacements of about 2000 m in the Chilean longitudinal valley during the Quaternary. In Japan the Kanto tectonic basin has subsided by as much as 1400 m in the Quaternary, whilst the Central Mountains have been uplifted by a maximum of about 1700 m. Highest rates of contemporary uplift include rates of 200 cm/1000 yr in the Kyushu Mountains. In general, half to two-thirds of the present height of the Japanese mountains is attributable to uplift during the Quaternary (Yoshikawa, 1974). In Turkey the very recent uplift may explain the apparent absence of early Pleistocene glaciations in its mountains. Similarly, in the Sierra Nevada, fossil floras in Pliocene and Pleistocene deposits are anomalously high. Reconstructions of their optimum climate may indicate subsequent elevation of some thousands of metres (Axelrod and Bailey, 1968).

Vertical movements may occur along faults, as in the Raton Basin in southern California where late Tertiary vertical displacements amount to at least 5 km of uplift. In fact, extensive normal faulting has continued in the entire Grand Canyon region since at least Miocene times. Landform interpretation and process studies must therefore reflect this simultaneous operation of endogenetic and exogenetic phenomena in the earth's many such active regions as in east-central Africa (Rossi, 1980).

ISOSTATIC LOADING AND REBOUND

Superimposed on the structural movements are elevations of land in areas relieved of the load of Pleistocene ice, affecting some 5 per cent of the earth's surface. Isostatic rebound is appropriately named for such movements exceed tectonic rates in many areas. Rates are difficult to compare since recovery diminishes rapidly with time (Fig. 19A). Any given rate approximately halves in an 800-year span, and rebound may be largely complete before a region is entirely ice-free. Rates of rebound, using the middle third of the emergence curve to obtain a representative value, include 29 m/1000 yr for Spitzbergen, 24 m/1000 yr for eastern Greenland, and 40 m/1000 yr for the west coast of Baffin Island. Maximum rates suggested include 70 m/1000 yr for Boston, which is similar to that for north-east Canada. In northern Scotland, a rate of 8.5 m/1000 yr may have occurred 12,000 years ago, slackening to 0.7 m/1000 yr in the last four millenia. The Gulf of Bothnia continues to rise at 10 m/1000 yr, with still one-third of the calculated total post-glacial rebound to come (Fig. 19B). Uplift so far is between 240 and 550 m. A recovery of 270 m may have occurred in Hudson Bay and a maximum depression of the central areas is possible.

Even on structurally stable coasts that are unaffected by postglacial rebound, the weight of water added to continental shelves may be sufficient to depress coastal areas isostatically in proportion to the average depth of water in the vicinity. The post-glacial submergence history of the eastern coast of the United States supports this hypothesis (Bloom, 1967). Sediment loading has a similar effect.

A further mechanism, 'driving subsidence', is now recognized at Atlantic-type margins, the trailing edges of drifting continents (Watts and Ryan, 1976). Such postrifting subsidence is attributed to increased density of the rifted basement, caused by thermal contraction or phase changes or both. Initially, the rate of subsidence at the shelf edge may be approximately 20 cm/1000 yr and, although the rate decreases exponentially over many millions of years, driving subsidence and associated sediment loading persists. Thus, on the edge of the continental shelf off eastern United States, no rate of subsidence of less than 2 cm/1000 yr has been measured in post-Jurassic times (Fox, Heezen and Johnson, 1970). This fact is geomorphologically significant since the Atlantic Coastal Plain was the area that prompted Davis's thoughts on the peneplain.

GLACIO-EUSTATIC CHANGES IN SEALEVEL

The total volume of present-day continental ice, most of which accumulated during the past ten million to twenty million years, is equivalent to 60 m of sealevel change. Thus, the growth and decrease of continental ice has probably dominated the sedimentary history of Atlantic-type margins since the mid-

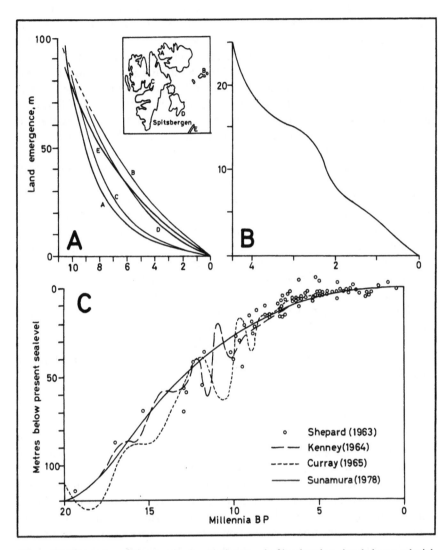

Figure 19 Examples of changes in the relative stand of land and sea levels in postglacial times

A   Isostatic crustal uplift in the Svalbard area, corrected for eustatic sealevel rise. The greater concavity of curves towards the south-east of the area may be the result of the weight of ice checking upwarping in the central area.

B   Land emergence at Stockholm, showing modification of isostatic uplift trend by eustatic factors.

C   Local variability and general trend of the world's postglacial sealevel changes, following the last glacial (Wisconsin or Devensian) maximum, about 20,000 yr BP.

Sources: A Grosswald, 1980; B Åse, 1980; C Sunamura, 1978.

Tertiary, some 20 million to 25 million years ago. Sealevel has fallen, intermittently and with fluctuations, since the Oligocene epoch. This trend has accompanied the refrigeration of Antarctica, which was associated with the opening of a passage south of Australia. Further cooling in the Middle Miocene followed, as an ocean current termed the 'South Ring Stream' found its way through the Drake Passage glaciers, beyond where the tip of South America now lies, thus further isolating Antarctica some 11 million to 14 million years ago. The considerable further cooling 5 million to 6 million years ago was due to tectonic uplift of several 100 m (Kvasov and Verbitsky, 1981). This accumulation of ice, in which Greenland also began to participate, accentuated sealevel fall, with lowering rate increased in the Plio-Pleistocene as ice sheets extended over North America and Eurasia (Tanner, 1968).

Sealevel stands during the Quaternary were shorter in duration and reversible. Although present levels are as high as might be expected in an interglacial, only 16,000 years BP the sea was probably some 140 m below its present level and possibly 30 m below this about 8000 BP. A few millenia ago, all ranges between two and six having been suggested, the present level was obtained (Fig. 19C). Evidence from the stable southern Florida area indicates that sealevel has not risen appreciably above its present stand in the last 4000 to 5000 years. Trends between 35,000 and 18,000 years ago are uncertain. Sealevel may have risen continuously from its low stand of approximately 140 m. Alternatively, Broecker concludes that sealevel fell between 25,000 and 18,000 years ago, using evidence from many sources, including former lake levels in the Great Basin.

TRANSGRESSIONS, REGRESSIONS AND STRANDLINES

The geomorphologist's appreciation of the nature and global significance of base-level change and stability depends greatly on the geologist's explanation of why shallow-sea shorelines advanced and retreated in the more recent geological epochs. For example, in the 1900s Suess and Chamberlin concluded that, when crustal movements occurred, they were upward on the continents and downward in the ocean basins, leading to a worldwide progressive relative elevation of the landsurface in relation to the sea. These views encouraged the supposition that the continents would carry, at increasingly higher altitudes, relicts of progressively earlier phases of denudational flattening, or 'cycles' of erosion.

Evidence now to hand suggests a number of causes that could explain why sediments of continental margins and cratonic basins often record transgressions and regressions that are synchronous over widely spread regions. In particular, a pulse of rapid sea-floor spreading could have substantially increased the volume of the world's ocean ridge systems (Hallam, 1971). This increase could account for the Late Cretaceous maximum transgression about 85 million years ago, which rose some 350 m higher than present sealevel, and submerged 35 to 38 per cent of North America (Hallam, 1977). The subsequent Cenozoic regression could be similarly explained by the later contraction of the mid-oceanic ridges. The estimated change in spreading rate from 2 cm/yr to 6 cm/yr is greater by a factor of two than rates of ocean volume change by any other mechanism (Pitman, 1978). Although the current explanation is totally removed from the ideas of the beginning of the present century, it suggests that base-level changes were closer to those assumed by the pioneer geomorphologists than subsequent geophysical evidence at first seemed to support. Peak transgressions were probably recognized worldwide at rates of 10 to 90 m rise/million yr, with more rapid falls of 95 to 170 m/million yr in regression (Hancock and Kauffman, 1979).

GEOGRAPHICAL TRENDS

The migration in space and in time of orogenic activity, rather than its occurrence as a sharp, synchronous, world-wide event, appears consistent with . . . plate tectonics. (Rona, 1973)

Rates and degrees of relative uplift may vary within relatively short distances. Axes of such warping may run approximately parallel or at right angles to coasts. As a well-known example of marginal downflexure of continents, the downwarping of the coastal plain and continental shelf of the eastern United States is documented in detail. This area, particularly in the margin of the Mississippi delta, includes the complication that, seaward of an hinge line, the sediments are downwarped whereas inland they have been uplifted. The Godavari delta in India has gradually prograded, but with conspicuous stillstands in the delta margin (Rao and Vaidyanadhan, 1979). Isostatic adjustment to sedimentary loading may be an important factor. Where warping axes are at right-angles to a coast, relict shoreline features may change level, due to warping in that direction. For example, there are three erosion surfaces in the coastal zone of Peru which span Pleistocene time, as indicated by deposits rich in a prolific molluscan fauna. These are the Mancora, Talera and Lobitos platforms, which are approximately 270, 150 and 20 m above sealevel at Cabo Blanco. However, when traced along the coast, levels change substantially. For instance, the Mancora platform falls 180 m in a distance of 72 km. Similar features are described on the central Baja California coast in Mexico, where the lowest terrace, the Tomatal, is some 120,000 years old and about 1 km wide. The Aeropuerto terrace is the highest, extends some 4.5 km inland, and inclines south-eastward at 0.2 m/km (Woods, 1980).

Geographical trend is a fundamental feature of isostatic rebound. Maximum rate of recovery may occur later in areas more centrally situated with respect to the ice cover than in more peripheral areas. Amounts of rebound inevitably tend to be largest where a former ice cover was thickest. Thus, uplift in Norway was greatest at the heads of deep fiords and least on the open coast.

The highest strandline marking the marine limit is at 220 m in the Oslo and Trondheim fiord areas. At places near the open coast, an uplift of only 10 m has occurred.

CONCLUSIONS

Because of glacio-eustatic changes in sealevel and the varied tectonic complications, sealevel positions before late Pleistocene can only be reconstructed where a wealth of varied evidence exists, and then these may be significant only to immediately adjacent areas. Even generally agreed indications of sealevel positions in earlier Pleistocene times are now realized to be much less certain than was widely anticipated twenty years ago. It would be extremely useful to have strandlines as global reference levels and some localities, like the Ben Lomond Mountain area on the Malibu coast, display flights of platforms that were undoubtedly cut by eustatically high seas (Bradley and Griggs, 1976). Stillstands occurred whenever uplift was cancelled by eustatic rise, and would thus characterize ends of periods of slowly rising sealevel. However, in practice, the occasional incontrovertible platform flights for a particular coast are rarely repeated elsewhere. Ironically, some of the most sharply defined platform flights owe the clarity of preservation to rapid uplift which itself then makes assessment of eustatic change difficult (Machida, Nakagawa and Pirazzoli, 1976); some have discernible gradients when followed along the coast. More generally, in the light of knowledge now accumulated, it seems unrealistic to separate base-level changes from pulses of erosion.

Structure, process and stage

INTRODUCTION

No phrase bears more repetition than W. M. Davis's statement that landforms are a function of structure, process and stage. However, its actual implementation into specific geomorphological studies requires careful scrutiny. In particular, any emphasis on just one element of this trilogy encourages self-supporting interpretations, as enquiries then converge on that element. Contrasts in the element singled out may account for many differences of opinion in geomorphology. Dichotomies are manufactured since exclusive attention to one element allows quite separate conceptions of the subject to run in parallel but without overlap. The distinctive emphasis of Davis's Cycle of Erosion was stage, with structural complications scarcely mentioned and much about process simply assumed. Even the structure of the present book, by introducing 'Process and Form' at an early stage (page 41) may suggest a weighting of emphasis away from geological stucture and landform stage.

An instructive, specific example is a process-orientated study in eastern

Jamaica. North-flowing streams, fed by uniformly distributed rainfall, have deep, round-floored meandering channels with sloping banks (Gupta, 1975). South-flowing streams have a contrasted wet-and-dry seasonal regime and are wide, shallow, flat-floored braided channels with steep banks. A map of these basins features in Fig. 2F. In addition to rainfall distribution, however, Quaternary crustal movements are a possible factor. With the north coast being emergent and the south coast submergent, progressive differences in gradient may be involved (Wood, 1976).

Two supplementary considerations might be added to structure, process and stage. The first, scale, was not apparent in Davis's generalizations, yet their relative importance depends on whether a single slope, subcontinent or grain of sand is being studied. The significance of scale is amplified by its tendency to be linked with time-span, since most smaller features are produced and modified more rapidly than larger forms. Secondly, the antecedent conditions envisaged are critical. These may differ greatly from the levelled peneplain, smooth veneer of sediments on an abruptly uplifted coastal plain, or other oversimplified points adopted by later workers. Not least, the antecedent conditions may be similar to those of today. Geomorphologists may underestimate how readily far-reaching but irrelevant conclusions are reached simply as a direct result of the basic assumptions adopted about antecedent conditions.

THE SIGNIFICANCE OF STRUCTURE

A comparison of the geological map of Ireland with a map of relief reveals a fundamental fact of Irish geomorphology: lithology and relief stand in the closest possible relationship. (Davies and Stephens, 1978)

Detailed studies often reveal that structure is an overriding control in certain areas. For instance, in the Shenandoah Valley, the geometric properties of drainage basins, stream courses and profiles, and other geomorphological features 'show a remarkable dependence on geological structure and on the distribution of rocks of different physical and chemical properties' (Hack, 1965). However, such localities in which structural control is a preponderant factor are not commonly sought in geomorphological enquiries, a tendency that invites some explanation. There are perhaps five possible reasons why the dynamic tectonic aspects of structure and the static lithological controls are not always stressed by geomorphologists. First, structurally controlled alignments and segments of the landsurface are ubiquitous. Yet here and there landform disregards structure, and it is these departures from structural control that have often intrigued the geomorphologist. Promoters of either process or stage emphases tend to regard structure as a minor, secondary or temporary influence to be erased by process and the passage of time. Secondly, structural control is often obvious. The interpretation of a structurally controlled landsurface shape may therefore seem dull, the prosaic pronouncement of the self-evident, dismissible in a couple of words. Thirdly, any suggestion of the importance of structure in valley development may appear to be a reversion to outdated notions, ever since Lyell (1833) pioneered Uniformitarianism by showing that rivers could erode valleys. Thus, on observing the control of stream courses in Australia by faults and recent warps and folds, Hills (1961) realized that he might appear to be taking a retrograde step in suggesting that these rivers were determined by crustal disturbances. Fourthly, continuous crustal movement is just not admissible when applying a Cycle of Erosion model. It also offends against ingrained Baconian thinking and its conditioning to prefer to consider one factor at a time.

Fifthly, and perhaps most fundamentally, geological influences could be overlooked as local complications before a coherent global model for earth structures was recently devised. The case of coasts can again be used, in this instance to exemplify the implications of plate tectonics for geomorphological generalizations. Previously, coastal classifications have followed W. D. Johnson's two principal categories of submerged and emerged coasts, apparently related to recent changes in sealevel. A secondary subdivision depended on whether erosional or depositional processes were predominant. More recently, Inman and Nordstrom (1971) suggest that the gross aspects of the relief of coastal zones, in the order of 1000 km or so in length, are related to their position on the moving plates of the seafloor. Collision coasts are all relatively straight and mountainous, with sea-cliffs, raised terraces and narrow continental shelves, typified by the west coast of South America. 'Trailing edges' are recognized on landmasses moving away from spreading centres, and another type occurs on the continental side of a marginal sea formed by an island arc. If the leading edge of such continents collides with another plate, the continued presence of mountains locates the largest drainage area on the trailing edge, notably the Mississippi and the Amazon. Thus, twenty-eight of the world's largest rivers, by drainage area, all discharge across trailing edges or marginal sea coasts. The Columbia River on the west coast of North America, only the twenty-ninth in world rank, is the largest to drain across a collision coast. Because of smaller drainage basins collision coasts receive less sediment than trailing-edge coasts, but the sand percentage is high due to greater relief. At scales of about 100 km, wave erosion predominates on collision coasts whereas depositional phenomena are more common along the trailing-edge coasts, and smaller water bodies favour delta formation.

Clearly Ollier and Pain (1978) demonstrate in their recent study of the structurally dominated geomorphology of Woodlark Island, Papua New Guinea, there is much challenge and new meaning in examining structural control. For more stable areas, as the case of karst geomorphology has always shown (Warwick, 1976), structure and process may be intrinsically inseparable.

THE REALITIES OF STAGE

The concept of 'stage' was possibly one of the most beguiling features of the Cycle of Erosion as it facilitated the drawing of the attractive analogy between landform change and Darwinian evolution of organisms. In the reaction against Davisian geomorphology, stage as its bastion has often been singled out for cross-examination, rebuke and even banishment. Admittedly, the Cycle greatly idealized and exaggerated the significance of sequential changes. However, many areas contain landforms resulting from more than one episode in the past to which the term 'stage' can be applied (Fig. 20). Equally, few geomorphological processes are intelligible without some context in time. This is literally the case in fluvial geomorphology where water level is referred to as 'stage', and spectacularly obvious during volcanic activity. Knowledge of the persistence and temporal changes in factors governing geomorphological processes has advanced strikingly in the past decade. Although commonly occurring simultaneously, three aspects of change in time can be identified. These are progressive trends, in which a change continues in the same direction over a span of time, oscillations about a mean condition, in which a certain regularity is discernible, and sporadic changes, which occur irregularly in time.

Progressive trends By far the most marked, persistent and far-reaching of the progressive trends are the inexorable horizontal shifts of continental drift, as the rearrangement of crustal segments continues to affect the evolution of the world's ocean circulations and climate. Continual equability prevailed back in the oceans and climate of Cretaceous times, but fragmentation of the world's oceans has led to an increasingly inefficient latitudinal exchange of energy. By the end of the Oligocene, some 22 million years ago, the ocean basins had assumed essentially their modern shape. While tropical surface temperatures remained stable throughout the Tertiary, high-latitude and, in consequence, deep-water temperatures continued to drop (Schnitker, 1980). Since the beginnings of Antarctic ice are now pushed back into the Cenozoic and as far back as the Eocene (Margolis and Kennett, 1971), a progressive if recently erratic fall in sealevel is a global trend. A 'sporadic' change may have acted as a trigger to initiate full-glacial conditions in the mid-Pliocence, some 3.2 million years ago, with the final closure of the Isthmus of Panama. This terminated low-latitude exchange of water and energy between the Atlantic Ocean and Pacific Ocean basins. The closure also increased the volume and velocity of the Gulf Stream some 3.8 million years ago (Kaneps, 1979). It is important to envisage landform evolution against a background of such environmental changes but without anticipating that some specific landform trait might be linked with them. Also, trends may be a distinctive feature of larger geomorphological systems too, which are independent of climatic change or structural movement, such as rivers in which a unidirectional migration persists. For example, high-level terrace deposits in the vicinity of the confluence of the Orange and Vaal rivers indicate that the Orange river migrated some 30 km northward during the Miocene and Pliocene (Helgren, 1979).

Oscillations about a mean condition   Superimposed on the longer term trends, there are a series of strict cycles of climatic change due to characteristics of the earth's solar orbit. These are eccentricity of the orbit, with an average period of 93,000 years, tilt of the axis between 22.1° and 24.5° over a 41,000-year average

---

Figure 20 Examples of stages in geomorphological histories

A   Re-excavation of a buried valley, the Milkweed Canyon 8 km from the edge of the Hualapai Plateau on the southern rim of the Grand Canyon.

Stage 1 A pre-tuff period of uplift and erosion exposed the Precambrian basement rocks, with the deposition of gravels of an ancient north-flowing drainage system which probably cross the line of the modern Colorado at a higher elevation.

Stage 2 Drainage disrupted by faulting and vulcanism and valley almost filled by localized flood gravels.

Stage 3 Deposition of Peach Springs Tuff about 17 million–18 million years BP, blocking all pre-existing drainage.

Stage 4 Re-excavation of the valley by knickpoint recession related to the gradual incision of the modern Colorado in middle and late Pliocene times, following regional collapse of the eastern edge of the Great Basin or uplift of the plateau.

B   Schematic river terrace sequences in the Upper Thames. The Plateau Drift may be related to fluvial activity in an early glacial stage, possibly Anglian. Compared with the local Jurassic materials of the Handborough Terrace, erratics suggest that the outwash source of the Wolvercote Terrace was some 30 km to the north, formed in the Wolstonian glacial stage. The lower portion of the Summertown-Radley Terrace is also probably Wolstonian in age, but the upper part is correlated with the last interglacial (Ipswichian). Downcutting of the buried channel occurred before gravels were deposited in the last glacial (Devensian).

C   Schematic marine terrace sequences on Bonaire, Netherlands Antilles. Although cut, and in part shaped by carbonate deposition, during a period of predominantly falling sealevel, each regression may have been followed by a transgression of lesser magnitude. Also, the solutional cavities reflect an intervening fall from the 70 m level to about 10 m before the later transgression to 50 m.

D   Moraines and lake shorelines left by the oscillating retreat of the last glaciation in western New York and southern Ontario. As Late Wisconsin ice retreated about 12,300 yr BP, early Lake Erie drained through 'Lake Tonawanda' in a multioutlet system. About 10,900 yr BP drainage focused only on the main Niagara Gorge as this enlarged. Of the several moraines named, L.E.–V.H. is an abbreviation for the Lake Escarpment-Valley Heads moraine.

E   Infilling of the Matumbulu reservoir, Tanzania, 1960–71. With an annual sediment yield of 729 m³/km², infilling at a rate of 13,200 m³ per year will be complete within thirty years.

F   Pattern of world increase in number of dams constructed. Distinctive stages in the pre-1900 and post-1945 periods are identifiable.

G   Mean annual decrease of ice thickness in recent years in the glacier Østerdalsisen, Norway.

Sources: Adapted from A Young and Brennan, 1974; B Sandford, 1954; C Bandoian and Murray, 1974; D Calkin and Brett, 1978; E Rapp *et. al.*, 1972; F Beaumont, 1978; G Knudsen and Theakstone, 1981.

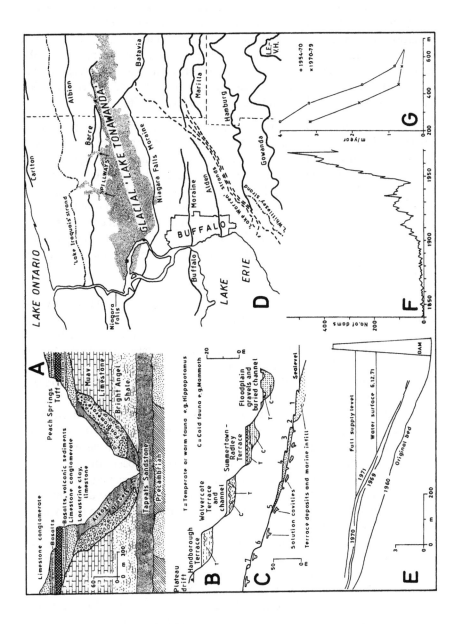

period, and the precession of the earth's axis, with a period of 21,000 years. According to the Milankovitch theory, ice ages are favoured when the total solar radiation for the half-year that contains spring and summer at certain northern latitudes is at a minimum (Weertman, 1976). The Milankovitch variation in solar radiation is similar to a north–south contrast in latitude of some 1000 km. Recently, the significance of the dominant periods in the earth's solar orbit is confirmed by ocean-floor sediments which suggest three dominant spectral peaks in iceberg deposits at 23,000, 42,000 and 100,000 years approximately (Hays, Imbrie and Shackleton, 1976).

Sporadic changes   Endogenetic changes are, by contrast, more sporadically distributed in time, as exemplified by episodic volcanic activity. For example, in the Guatemalan Highlands, reverse grading and many tephra layers probably indicate a progressive increase with time in the intensity of eruptions. Other layers have alternating fine- and coarse-textured horizons, suggesting cyclic change in the intensity of eruptions. The thickness and profile development of paleosols show that durations of quiescence between eruptions varied (Koch and McLean, 1975). The broader effects of such effects include the now-convincing evidence for simultaneous occurence of vulcanicity and glacial events (Bray 1974).

Although many early geomorphological episodes took place against a background of climatic, astronomical and structural change, it is only within the Quaternary era that stages begin to be recognizable with some certainty. It is scarcely surprising that the Quaternary is the age of stages. Within its climatic and associated sealevel oscillations, innumerable stratigraphical relationships record the contrasting stages of landsurface evolution that occurred. Indeed, for sediments offshore, the term 'palimpsest' has been introduced to describe reworked Quaternary sediments with textural characteristics reflecting an earlier depositional episode and also a subsequent regime (Swift, 1971). In river valleys, relict forms such as terraces record sequential developments. Even the 'morphoclimatically stable' central Sahara has been affected where a widespread depositional surface, the 'Middle Terrace', accumulated progressively between c. 17,500 BP and c. 6,600 BP when there was a sudden influx of pebbles. This change was possibly related to a transition between two types of wetter periods, that of more frequent tropical depressions giving way to monsoonal rains of the Sahelian type (Maley, 1977). Even channel cross-section shape changes over short periods of time, recognizable in terms of distinct phases (Knighton, 1975). Ultimately, the most emphatic stages are those of ice itself. Although deglaciation of the British Isles was complete by about 12,500 BP, the Loch Lomond Readvance about 10,800 BP was sufficient to leave vestiges of this substage at over a hundred locations (Sissons, 1974).

MANKIND AND OUR TECHNOLOGY AS A GEOMORPHOLOGICAL STAGE

Few stages in landform development and change are as much in evidence as those directed or unintended by the human race. The Roman Empire offers innumerable examples of both possibilities. For example, in the Lago di Monterosi area 40 km north of Rome, the construction of the Via Cassia through the area increased erosion rates from 2–3 cm/1000 yr to about 20 cm/1000 yr. Man's influence can be seen today in many suburban areas throughout the world, where constructional activity is inducing several rapidly changing stages of landform. These include basal slope undercutting, due to floodplain expansion following increased sediment transport, and deposition (Graf, 1976).

From the draining of the Fens (Darby, 1940) to the Aswan Dam, anthropogenic changes introduce significant changes in fluvial processes, sediments and forms. Dam construction markedly alters flood-frequency distribution and the resultant flows may fail to entrain debris supplied to the main channel from tributaries (Petts, 1979). For example, in the six years after the construction of the Gardiner Dam on the South Saskatchewan river, median discharge was nearly halved and summer flows markedly reduced. Mean grain size increased from 0.53 mm to 0.67 mm, due to fines being winnowed, leaving an armour of coarser materials, which cannot be transported (Rasid, 1979). Many changes accelerate sediment yields, such as the excavation of drainage ditches, as monitored in central Wales (Newson, 1980).

Such man-induced fluvial adjustments are an increasingly absorbing challenge for students of river systems (Park, 1981); every year new instances are introduced and the length of records of earlier changes already being monitored is extended. The implications for geographers are also marked, due to the substantial regional variation in the scale of anthropogenic impact. On the broadest scales, some 20 per cent of the total runoff in North America and Africa is regulated by reservoirs compared with only 4 per cent in South America and 6 per cent in Australasia (Beaumont, 1978). In the case of Japan nearly all rivers have been modified by man, and the hydrological character strongly affected. Most recently, reduction of sediment supply to coasts, both because of dam construction and river-bed mining, has caused coastal retreat in the last two decades (Ota and Nogami, 1979).

The categorization of mankind and our technology as a stage in landform development is convenient, as it accommodates human activities within the scope of a natural science without clouding its objectivity with narcissism. To classify human activity as one element of the fourth dimension (see Table 1) is also quite logical and appears to narrow the separation between the poles of two dichotomies. First, the classification of man as a stage in time works equally effectively whether the human race is seen as the latest link in an evolutionary chain of geological span, or whether mankind is the relatively recent, special creation of a deity. Secondly, any distinction between Man and Father Nature appears to be irrelevant if man is seen as part of time. Also, studies of man-induced landform stages lends coherence to geography, as such studies interdigitate with, if they do not become an integral part of, historical geography.

TIME-INDEPENDENT LANDSURFACE FORMS

Two separate lines of argument arrive at the concept of time-independent landforms, both deliberately conceived to expose the alleged weaknesses in the emphasis on sequential change with time in Davisian approaches. The problem of how these arguments can coexist with the evidence of the reality of stages at hand is puzzling. First, Hack (1960) has revived the concept of dynamic equilibrium introduced from physics to landform study in the late nineteenth century by G. K. Gilbert, who believed in a geomorphological world oscillating in time and space. It replaces directional concepts of geological time with a 'steady state' in which work done and imposed load are approximately in balance, outflows equal inflows, and forms remain unchanged. Thus the ridge-and-ravine of the Appalachian-type area suggested a quasi-equilibrium between stream-channel downcutting and the erosion of slope and divide. This is a meaningful interpretation of the type area. However, the concept is rather narrow and difficult to apply to other areas. It offers no explanation for relict features on broad interfluves, as in the Ozarks, where present-day weathering is essentially independent of stream-channel activity. It offers no alternative to the Cycle of Erosion in that, even if landforms remain essentially the same over long periods, downwearing will still tend to approach some base-level. In fact, sediment accumulated offshore from the Appalachian ridge-and-ravine area show that recent rates are about one-eighth of those of post-Triassic times (Menard, 1961), a classic example of the winding down of the effectiveness of processes over time. Clearly, the concept of dynamic equilibrium does not necessarily imply steady state.

In the 1930s both Ashley and Fenneman had described 'non-cyclic' erosion causing the straight, horizontal Appalachian crests to be lowered some hundreds of metres, yet looking the same afterwards as before. Their interpretations were based on the structural control of landform development due to the parallel belts of rocks with differing resistance to erosion. The visual impression does indeed support the analogy with dynamic equilibrium. The explanation of these phenomena, however, is probably structural control. In fact, in the southern Appalachians, Hack himself prefers structural control in attributing domed topography in crystalline rocks to sheeting.

A second line of reasoning has supported the notion that landforms may be time-independent. Drainage-basin networks in homogeneous rocks reveal close correlation between parameters such as stream numbers, lengths and the frequency of confluences between tributaries. These correlations exemplify Horton's 'laws' by demonstrating unvarying properties in drainage basins

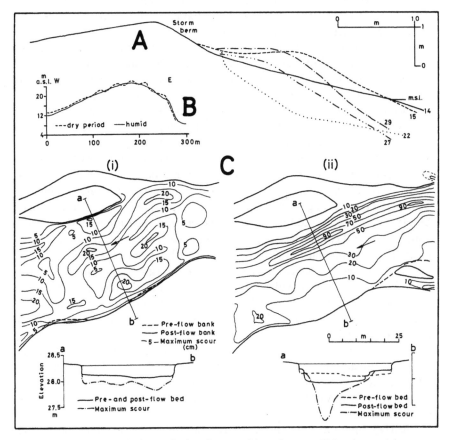

Figure 21 Oscillations in forms during the reworking of unconsolidated materials

A   Beach profiles at Half Way Road, Delaware. The numbers on the profiles indicate the dates of measurements in June 1976. The sequence shows the storm berm created prior to 14 June, with new tidal berm and beach cusps constructed on the following day. These were then slowly obliterated. On 22 June waves eroded the face of the tidal berm followed by deposition and progradation of a tidal berm by 29 June.

B   Long profile of a barchan dune in Słowinski National Park, Poland. Migratory mesobarchans 8–60 m long and 0.3–2 m high develop in dry periods on the surface of the main dune. Accumulation of material on the distal slope of the main dune is considerably *accelerated* during humid periods because airflow drag close to the ground is much reduced by the elimination of the mesobarchans.

C   Maximum scour after floods in Quatal Creek, Ventura County, California. (i) January 1974, (ii) December 1974. The maximum scour contoured is the difference between the pre-existing bed elevation and the elevation of the base of the reworked material. The cross-sections show that these active channels are stable in moderate floods.

Sources: Adapted from A Dubois, 1978; B Borówka, 1980; C Foley, 1978.

which are therefore independent of the passage and changes of time. They reflect an optimal spatial organization for drainage which, should it cease to be functionally related to a changed climate, would be re-established with the same morphometric attributes following renewed fluvial activity (Ongley, 1974). This invariability of pattern is again, fundamentally, a distinctive type of geological control. The very lack of structural or lithological guidance to drainage in a homogeneous rock outcrop is a dominant influence in allowing surface runoff to drain downslope in directions determined purely by chance. The stream parameters and their interrelationships are therefore free to obey the stochastic laws of averages, chance and probability.

There are two further categories of time-independent landsurface forms, each representing the opposite extremes. On the one hand, a landsurface may be so resistant that it remains unmodified by subaerial exposure throughout protracted periods of geological time. For example, the surface of the Canadian Shield has been lowered little, and the pattern and appearance of the bedrock landsurface and landsurface configurations are essentially expressions of bare-rock structure (Godard, 1979), unchanged at least since the Ordovician. In South Australia, the laterite surface of the Gulfs region has persisted through some 200 million years of subaerial weathering. Similarly, the forms preserved on the granite residuals of the Eyre Basin have withstood many millions of years of exposure (Twidale, 1976). At the opposite extreme are landsurfaces that offer virtually no resistance to movement (Fig. 21). Here, the structural control is the very incoherence of materials which are then refashioned readily by every ebb and flow, thread of current, or brush of wind. Forms such as the barchan dune are fashioned much in accordance with physical laws, are readily interpreted as physical systems, and their shapes are nearly as old as the Canadian Shield.

Clearly, time-independent landforms are as much a geomorphological reality as stages. The puzzle of their coexistence is very simply solved by utilizing Davis's simple but fundamental formula. Each of the four categories of time-independent forms is, in contrasting ways, a function of structure. If a landform is essentially a function of structure, it follows that it will be essentially independent of time.

The necessity for simplification of geomorphological complexity

INTRODUCTION

It is difficult to grasp the nature of phenomena that are more than moderately complicated, whether the complexity be verbal, statistical, temporal or spatial. Geomorphology tends to encompass all four. Some recent work, particularly in stream-channel morphology, draws much inspiration from Gilbert's work, and some justification from his conclusion that 'the development of complexity within complexity suggests that the actual nature of the relation is too involved for disentanglement by empiric methods' (Gilbert, 1914). Conversely, theories are not sufficiently comprehensive to predict the dimensions and shape of many landforms. For instance, in the case of sediment banks and shoals in erodible-bed tidal estuaries,

> such a shortcoming is understandable in view of the hydraulic complexity of tidal entrances. Tides, wind, waves and swell, Coriolis effects, vertical and horizontal density gradients, fresh water runoff, and unsteady nonuniform nature of the flow combine to frustrate exact analysis. (Ludwick, 1974)

If added to such variables are tectonic movements, contrasted lithologies, climatic changes, impacts of escalating technologies and irregularities of most landsurface shapes, some manner or mechanism for simplification in geomorphology seems essential. Conversely, to ignore complexity is to turn a blind eye to a fundamental aspect of reality. To attend in detail to the individual case may be to lack a vision for broader relevance. There appear to be two contrasted bearings for steering a course between the Sirens of excessive simplification and of superfluous detail, the approaches of conceptual modelling and of regional geography. Both are, in part, dependent on the laborious task of empirical classification.

CLASSIFICATION

Information must be ordered if it is to be digested, as commonly achieved by classifying various observations into categories. Classifications are usually arbitrary systems designed to facilitate handling a range of phenomena at the risk of some distortion of the truth (Sparks, 1971). The more differentiating characteristics included in the scheme, the closer the more complicated classification approaches reality. Successful classifications usually depend on there being certain distinct types of natural phenomena. In geomorphology, continuous gradation is as likely as discrete categorization.

After the lull following the collapse of genetic classifications, some tentative morphological classifications are beginning to reappear for forms for which a bulk of data is to hand or has been specifically collected. Fluvial-channel attributes, in particular, are increasingly classified (Fig. 22). A thoroughly substantiated classification of hillslope morphometry in New Zealand revealed that length/size and shape attributes were most significant (Blong, 1975). However, a study of coastal types in Venezuela concludes that no really satisfactory classification has yet been established which characterizes both landforms and coastal dynamics (Ellenberg, 1978).

Regional variations in landform are such that classification probably still has most relevance at the local level. Since successful classification depends on the amassing of a sufficient bulk of data, the degree to which classification reemerges in geomorphological work may be some index of the status of accumulated measurements in that branch of landform study.

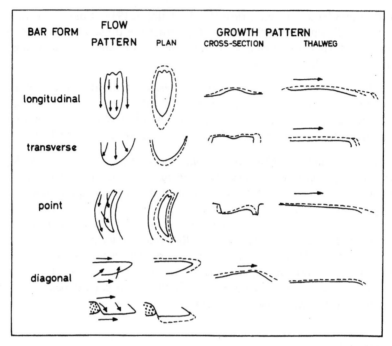

Figure 22 Example of a process-orientated classification of landforms–generalized flow and growth patterns for stream-channel bar forms. Dashed lines indicate accretion.
Longitudinal bars, favoured by wide, shallow channels where banks do not affect flow.
Transverse bars, with flow lines diverging by channel widening, abrupt depth increase, or at junctions of two channels of unequal flow strength.
Point bars, on curved channels with erodible outer banks.
Diagonal bars, where channel cross-sectional flow is asymmetrically distributed.
Source: Adapted from Smith, 1974.

CONCEPTUAL MODELS

> *The height of folly is to indulge in wishful thinking and fail to face reality.*
> *(J. B. Boussuet)*
> *I live on the ninety-ninth floor of my block, imagining the world outside.*
> *(M. Jagger and K. Richard)*

As a substitute to the labour of collecting data, a conceptual model may be erected instead. Ideally, the two go hand-in-hand, as the classic model-builder, W. M. Davis, demonstrated in structuring much of the thinking and fact-finding of the nineteenth century into his Cycle-of-Erosion model. His achievement remains an excellent example of method, since any internally consistent conceptual model must incorporate all relevant field observations and experience, together with appropriate experimental data. The sequential aspect to Davis's view is now recognized as a 'Markovian property' where present landsurface features are dependent on those that precede them either in space or in time. The purpose of modelling is to test for compatibility between a model and new observations and empirical data. Thus, in stream-channel geomorphology, 'actual field examples should form a kind of statistical array that peaks somewhere near the theoretical value' (Langbein and Leopold, 1964). If new data contradicts the model, this merely stimulates further experiments, field observations and, in particular, a search for additional variables. Thus, in the streams studied by Fisk (1977), several simple dilution models failed to predict the variation in solute concentration with discharge in all but one actual stream. The importance of additional variables, therefore, had to be postulated, such as seasonal variations in groundwater flow and bank storage in flood plains. If new data do not contradict the model, this does not verify the model's validity. Even if field data can be fitted into a model, this does not prove the model adequate either, because the same data might fit other schemes equally well. Thus, in simulating stream headwater growth and branching, Howard (1970) concluded that all models he tested were equally good approximations. This merely suggests, however, that only a superficial resemblance had been established. A successful model must prompt a satisfactory interpretation of actual cases. Thus, geologists who investigate naturally folded strata are satisfied that models and natural processes correspond extremely closely.

To ensure internal consistency in structuring a conceptual model, systems theory and analysis is used to identify functional units and controlling processes within the structure of complicated interactions. This approach is based on general principles that apply to patterns of knowledge as a whole. These are therefore recognizable in any specialist branch of knowledge, such as geomorphology (Strahler, 1950), physical geography (Chorley and Kennedy, 1971), or geography as a whole (Chorley and Bennett, 1978).

In geomorphology, equilibrium is achieved in 'open' systems if arrival of one material equals the rate of escape of another. For instance, as described by W. M. Davis, the outflow of waste material from a hollow on a slope may balance that of inflow. Where this state of balance or equilibrium can be postulated, systems analysis describes a 'steady state'. However, even in the highly dynamic beach environment where easily reworked unconsolidated materials are involved, inherent instability in wave and sediment parameters permits only a modified form of steady-state conditions. If parameters fluctuate about a continually shifting condition, 'dynamic equilibrium' is then recognized. However, at any point in geomorphological time, one set of forces is usually stronger than others, so that a sequence of events is discernible rather than equilibrium (Fig. 21). The distinction usually depends on the time-span over which conditions are considered. Despite appearances of balance on beaches, even in the short span of recent years there may be a decrease of sand supply in relation to amounts removed down submarine canyons.

Some geomorphologists are very uneasy about 'dynamic equilibrium', since usage of the term does not make geomorphology intrinsically any more scientific nor facilitate the elucidation of fundamental principles. Ollier (1968) advises that 'the danger of confusing the geomorphic 'dynamic equilibrium' with chemical dynamic equilibrium must be constantly borne in mind lest we are deceived by our own metaphor'.

'Feedback' mechanisms describe modes of change (King, 1970). These include situations where processes are, up to a point, self-enhancing (positive feedback) or self-regulating (negative feedback). A clear example of a self-enhancing process is the drying that produces contraction fissures in clays and soils. Once initiated, a fissure then promotes the drying and aeration of deeper horizons, and the fissure deepens. Eventually, a complicated set of self-arresting processes limits the depth of the fissures, and, although one might then expect some quasi-equilibrium to develop between these two sets of processes, the dry spell might well have ended before this condition was approached. A good example of a self-arresting process is that which limits the height of sand dunes. As these become higher, the greater the contrast becomes between wind speeds on the dunes and over the troughs. Therefore, the higher the dune the greater the tendency for sand to be carried over the dune crest and deposited in the troughs. A more complicated example is P. Pinchemel's concept of 'auto-assechement' in Chalk dry valleys. Identification of feedback loops can sometimes epitomize the essential difference between interpretations. For instance, headward erosion in W. M. Davis's scheme of drainage development was a self-accentuating mechanism, a 'positive feedback loop' leading ultimately to river capture. In contrast, in an alternative interpretation of scarpland drainage evolution headward erosion is regarded as an ineffective, self-arresting process (Pitty, 1965). Data recently acquired near Denver prompted the conclusions that

> As the gully length increases, the headcut retreats further into the headwater area of the system where lesser discharges and lower amounts of energy are available for erosion, so that increasing length dictates decreasing erosion rates. The result of this self-regulating negative feedback loop is that migration rates for the headcuts are eventually reduced nearly to zero. (Graf, 1977)

Where all relevant data are available and an internally consistent system with definable boundaries has been identified, the conceptual model might be constructed in mathematical terms. The main hypothetical reservation about this approach is that the many idealizations needed are usually important only for mathematical tractability, essential for mathematics but not necessarily so for the geomorphological object being examined. Conversely, idealized assumptions are not necessarily made about those aspects of the real world that are most susceptible to simplification. For example, some mathematical models of stream-long profiles can be formulated only if 'complications' such as tributaries, are absent or ignored. In addition, the very incompleteness of geomorphological knowledge, especially its historical and geological framework, makes the definition of boundary conditions notionary, and complete explanations unobtainable. In practical terms, the prerequisite data are often lacking and sometimes unobtainable. Thus, many of the mathematical models that have been proposed over the last century are simply sophisticated assertions about the unobservable. Similar conclusions have also been reached in other aspects of geography and in methodological developments in the field of economics, with some models becoming so far removed from reality that they are, in the last analysis, immaterial (Floyd and O'Brien, 1976).

In extreme cases, mathematicians may even begin with a comprehensive law and try to deduce geomorphologically significant principles and generalizations from it. Persistence with this approach depends on a willingness to accept discrepancies between the law and the reality without seriously doubting the validity of either. In geomorphology, this credibility gap is often quite wide. For instance, it has been suggested that pyroclastic flow deposits will conform to Rosin's Law of Crushing, in which particle size decreases exponentially. However, in studies of the El Pajal deposits in central America, not one of the particle-size distributions analysed followed Rosin's Law (Davies, Quearry and Bonis, 1978). This approach is most frequently encountered in stream-channel geomorphology. However, attempts to explain the shape of the average meander bend, or even the tendency toward bend development by application of the laws of statistical thermodynamics, must be viewed sceptically because of randomness and difficulties in defining and discussing it (Hooke, 1975). In other instances, law and theory can be confused if a law is applied empirically without examining the nature of the system. Perhaps the most disconcerting simplification is that geological heterogeneities can be overlooked or that they can be dismissed as aberrations. This is evident in the classic statement on channel links where the study area 'is underlain by flat-lying Pennsylvanian sandstone and shale, essentially free of geologic controls that might distort the almost classical dendritic pattern of the stream network' (James and Krumbein, 1969).

REGIONAL GEOGRAPHY

Most of us have come to accept the principle that geomorphic hypotheses and their applications vary with the geographical settings of areas. (Tuttle, 1975)

Knowledge of regional geography simplifies landform studies in many ways. One is the use of information from the classic area where a given phenomenon is particularly well displayed. For instance, information of global significance on the interrelationships between tectonic movement and relief, and their influence on erosion and deposition, comes from detailed studies of present-day movements in California. Similarly, all the phenomena of processes at colliding plate boundaries in island arcs have been studied extensively in Japan. Another example: 'Iceland is geologically unique in several ways. It is the world's largest mid-oceanic island and is the only segment of the mid-oceanic ridge system that is well exposed and clearly and actively spreading' (Watkins and Walker, 1977). Iceland is also the classical land of outwash plains and also for waterfalls of very varied origins.

There is some distinction between the 'classic' area and the type area. The significance of the latter is partly due to the intensity of existing investigations often located in areas where certain features were first recognized. Thus, possibly the world's greatest concentration of pingos lies in the Mackenzie delta area, and Spitzbergen may be the classical land of the soil polygon, but it is in Poland that Quaternary periglacial features are perhaps best known. A very important category of type area is that where much information has become available as a byproduct of other endeavours: the long history of geological investigations for the rich economic resources of the Rhine graben has provided extensive, little-known data on rift structures. For different reasons, the best marine platform profiles on the Malibu coast of California were on the well-preserved Santa Cruz platform, in part because 'Brussels sprouts were grown there. Farm roads proved to be ideal for seismic surveys because they were flat, well compacted, free of vegetation and oriented parallel or perpendicular to the shoreline' (Bradley and Griggs, 1976).

With changing interpretations, the features of some classic or type areas may become of historic interest only. For instance, in 1947 Western Australia was still regarded as a natural eustatic gauge on account of its assumed structural stability; evidence now suggests that there were late-Tertiary epeirogenic warping and major folding movements in the Pleistocene, and that earthquakes in recent years occurred along what is now believed to be the remaining side of a now-vanished rift valley.

In some geographical situations, there is a natural control which eliminates a variable, as shown by comparisons between lake and coastal shores. The absence of tides and smaller-size waves on lake shores means that terraces are better formed and deltas are simpler. Mixing of saline and fresh waters is usually absent and shore zonation lacks the complexity of tidal areas. In the Nile, unlike most streams, there are no 'complications' to downstream progressions because little water and sediment enters the river during its long traverse of the desert area. However, the Aswan Dams have substantially modified this progression and, increasingly, in terms of deliberate choice of including or discounting a natural control in geomorphology, human activities loom large. Geologists, in particular, are concerned with areas where human activity is minimal, if erosion rates calculated are to be related to the geological column. Also, the same studies are vital for geographers, if only to set a semi-natural threshold in the past from which to scale the subsequent effects of human activities. Tactful use of terminology, however, is vital, as to 'eliminate' man can be misinterpreted as narrowness if not sacrilege. Similarly, the advantage of areas 'free' from vegetation must be described in terms which do not appear to devalue the importance of ecological studies.

A particularly valuable natural control, relating to landform itself, is found in areas where some natural geometrical simplification of complexity can be found. For instance, Huntley and Bowen (1975) chose Start Bay, south Devon, for a preliminary experiment to measure the nearshore velocity in waves. Here, Slapton Beach has 'a relatively simple topography along a fairly straight coastline'. Similar arguments are advanced for slope measurements in areas of rectilinear contours (Pitty, 1966). A much-used natural geometrical control is scale, as simpler as well as more manageable topics are identified. For example, the small size and slow rate of movement are typical characteristics of many cirque and valley-head glaciers in Scandinavia. Elsewhere, deformation in much larger valley glaciers obliterates ice structures. Hillslope studies by Schumm (1956) concentrated on manageable, smaller-scale landforms in essentially soil-free areas. At the smallest scales, sedimentologists may seek out areas where variations in grain size are minimal, in order to investigate the effect of grain shape.

Complementary to natural controls like these is evidence of the same factor or feature being present in quite different environments. For instance, barchan dunes have been noted on deep-sea floors. Meandering currents in the Gulf Stream and on ice (Ferguson, 1973) are very similar in appearance to river meanders. The significant factor is that ice channels or ocean currents exhibit meanders in the absence of sediment load. The importance of an abrasive in the development of stream meanders might, therefore, be minimized.

Certain rocks introduce important simplifications into the study of land-forms. Solutional studies of limestone are favoured by the soluble, mono-mineralic composition of the rock. In tropical areas limestones offer the best opportunities for studying the recession of cliffed coasts. A conglomerate offers an unusual opportunity for the study of rock weathering because all conditions, other than the parent material provided by the individual frag-ments, are equal. Areas of unconsolidated sediments are particularly prized by investigators wishing to apply systems analysis and mathematical models to their work. Without the necessity for a preliminary weathering phase, water easily reworks unconsolidated material, and regularity of form and balance of process quickly develop or reform.

Distinctive climates may introduce some control. For instance, in the study of wind action on sand, the absence of vegetation was an advantage compared with coastal dunes (Bagnold, 1941). Identification of structural lineaments and their associated geomorphological patterns is favoured by the characteristics of the arctic environment, such as absence of vegetation, low coastal energy, and perhaps the presence of permafrost (Short and Wright, 1974). Perhaps the most vital natural 'control' for geomorphology is seasonal change in weather. Rather than encouraging a predominantly chronology-orientated study, such changes are regarded as offering a range of natural inputs into real geomor-phological systems, the environmental equivalent of changing the setting on a laboratory dial.

A science is commonly recognized by statements about its subject matter which can be substantiated by repeated observation. Often, only a very comprehensive knowledge of regional geography can identify or predict where areas with sufficient geomorphological similarity might be found. If it is part of the role of a science to formulate the conditions under which a situation repeats, the geographical element in geomorphology is essential in identifying the locations where a given situation might recur.

Glossary

altiplanation. The combined effect of a range of geomorphological processes in producing sub-horizontal benches and associated steps in periglacial environments.

asthenosphere. The plastic layer in which magma is generated some 50 – 240 km beneath the overlying rigid lithosphere of the earth. Isostatic adjustments and lateral movements of lithospheric blocks and plates take place within this zone.

barchan. A Turkestan word for an isolated crescent-shaped sand dune. The upwind surface is convex in profile with an oval edge in plan. Two 'horns' extend downwind on either side of a hollow which lies at the base of a steep slip face.

batholith. An intrusive igneous rock mass, usually of granitic material. It has an irregular upper surface which is discordant with the disposition of surrounding rocks and has no discernible base.

berm. An ephemeral, sharp-fronted ledge of beach material thrown up by storm waves.

biomass. The weight or volume of organisms in a given area.

BP. An index of time 'before the present', usually taken as AD 1950. This age is commonly established by radiometric dating, particularly by the carbon-14 method.

caldera. A large, basin-shaped depression within a near-circular rim of volcanic materials, usually attributed to the subsidence or explosion of a volcano.

Carboniferous. The last geological period of the Paleozoic era, following the Devonian and preceding the Permian. It began about 370 million years ago and lasted for nearly 100 million years.

Cenozoic. The latest geological era, following the Mesozoic and extending to the present. It began about 70 million years ago and includes the Tertiary and Quaternary periods.

cirque. A half-bowl-shaped rock basin with a striated and smoothed concave floor, commonly found at the head, and high on the sides, of glaciated valleys.

clastic material. A transported sediment consisting of fragments of comminuted rock (clasts) and detached minerals.

continental drift. A hotly disputed theory when originally put forward by A. Wegener in 1912, which accounted for the distribution of continents and ocean basins in terms of break-up and lateral drift of the continental crust. Currently, such displacements are accommodated within the plate tectonic model with its more clearly substantiated mechanisms, such as sea-floor spreading.

Coriolis effect. The deflection of moving particles due to the centrifugal force created by the earth's rotation.

craton. An ancient, essentially immobile and non-volcanic portion of a continent. Cratons may incorporate broad sedimentary plains which overlap the crystalline shield areas at their core.

Cretaceous. The final geological period of the Mesozoic era, following the Jurassic and preceding the Tertiary periods. It began about 135 million years ago and lasted for some 65 million years.

cusps. Regularly spaced, crescent-shaped accumulations of coarser beach material, with their 'horns' extending seaward.

denudation. The combined effect of geomorphological processes involved in rock weathering and debris transportation. In its original usage, the exposure of bedrock was implied.

Devonian. The first geological period of the Upper Paleozoic era, following the Silurian and preceding the Carboniferous. It began about 400 million years ago and lasted for some 50 million years.

diastrophism. The combined, large-scale effects of all endogenetic processes which deform the earth's crust.

diorite. A coarse-grained igneous rock containing a proportion of minerals intermediate in type between those of acidic and basic igneous rocks. It is composed essentially of plagioclase feldspar and dark-coloured hornblende. In contrast to granite, only small amounts of quartz may be present.

distal. An adjective commonly used in discussing geomorphological transportation, describing those sediments and forms furthest from a source area or point of onset of movement.

downwarp. The regional subsidence or bending down of a portion of the earth's crust, characteristic of some coastal plains such as the Atlantic

Coastal Plain of North America.

endogenetic. An adjective describing the geophysical and geochemical processes operating within the earth's interior, commonly used in contrast with the exogenetic processes of geomorphology operating at the earth's land surface.

environmental determinism. A notion, dating back to Hippocrates, that cultural differences may be caused by regional differences in the natural environment. Climate, in particular, has been singled out for emphasis, as by Montesquieu in 1748. This form of determinism was briefly revived for geography at the beginning of the twentieth century by Friedrich Ratzel's criticism of Darwin and by the influential writings of his American student, Ellen Semple, and by the prolific work of Ellsworth Huntington.

Eocene. A geological epoch of the Lower Tertiary period, following the Paleocene and preceding the Oligocene. It began about 55 million years ago and lasted some 35 million years.

epeirogeny. The essentially vertical component in diastrophism.

eustatic. An adjective describing worldwide changes in sealevel, particularly those of the last few million years caused by changes in the volume of continental ice-caps. Sea-floor spreading is increasingly being considered as a possible cause of earlier eustatic changes.

evapotranspiration. The vapour transfer of moisture to the atmosphere by transpiration of organisms and by evaporation from the land surface.

foraminifera. Types of minute planktonic organisms with many-chambered shells of calcium carbonate which are deposited as organic oozes.

geosyncline. An obsolescent nineteenth-century term for an elongated downwarp in the earth's crust in which huge thicknesses of sediment accumulated before uplift and created mountain chains.

geothermal. An adjective describing the heat of the earth's interior.

gilgai. Patterned microrelief of expanding clay soil in lower-latitude regions with marked seasonal change in soil moisture.

granodiorite. A coarse-grained plutonic rock, intermediate in composition between granite and quartz diorite, with plagioclase the predominant feldspar and with some quartz.

greywacke. A hard, dark-coloured sandstone composed of clasts and easily weathered minerals embedded in a much finer matrix.

Hercynian orogeny. The late-Paleozoic mountain-building era of the Carboniferous and Permian periods, characterized in Europe by a north-westerly fold trend.

holism. A philosophical term describing how methodological collectivists treat large-scale bodies of knowledge as 'wholes'. In geography, the term is much used to indicate the integration of physical and human components, as opposed to the individualism of loosely bound aggregates of specialist knowledge.

Holocene. The later geological epoch of the Quaternary period, following the main Pleistocene glacial episodes and extending to the present. So far, it has lasted about 10,000 years. This epoch, therefore, is the 'post-glacial' of mid-latitudes and is known as the Recent in many discussions.

idiographic. A little-used adjective, employed in geography to emphasize studies in which cases or areas are treated individually, with a view to understanding distinctiveness of each one separately.

isostasy. The tendency for the brittle segments of the earth's crust to maintain a near-equilibrium balance, adjusting to unloading and loading effects of denudation, sedimentation, and ice-volume change, as they 'float' on the underlying, denser and plastic asthenosphere.

Jurassic. The middle geological period of the Mesozoic era, following the Triassic and preceding the Cretaceous. It began about 200 million years ago and lasted for about 65 million years.

kaolinite. An alumino-silicate clay mineral, typical of acid weathering environments. It has a very limited capacity for base exchange and does not expand with increased moisture content.

karst. A type of terrain recognized where limestone solutional forms are well developed, as in type area of the Yugoslavian Karst. Characteristic forms include enclosed depressions and sinkholes, separated by irregular ridges. Rock outcrops are prominent, and dry valleys and underground drainage are common.

knickpoint. An abrupt increase of gradient in the longitudinal profile of a stream, either linked to the outcrop of resistant strata or marking the current upstream limit of increased channel incision downstream.

Laramide orogeny. A mountain-building era extending from the late Cretaceous to the end of the Paleocene, for which the Eastern Rocky Mountains are the type area.

lithology. The general physical character of a rock.

lithosphere. The rigid, outermost layer of the earth, made up predominantly of solid rocks.

magma. The molten, mobile rock material, generated below the earth's crust, which gives rise to igneous rocks when solidified on intrusion into the crust or when extruded on to the earth's surface.

Markov process. A statistical process in which, in a sequence of random events, the probability of any event is influenced by the outcome of the preceding event.

Mesozoic. A geological era, following the Paleozoic and preceding the Cenozoic. It began between 270 and 230 million years ago, lasted until some 65 million years ago, and comprises the Triassic, Jurassic and Cretaceous periods.

montmorillonite. An alumino-silicate clay mineral, produced under alkaline conditions from alteration products of basic rocks. It has an expanding-lattice crystal structure and can thus adsorb large quantities of interlayer water, swelling considerably in the process.

morphogenesis. A term common in climatic geomorphology, usually denoting the distinctive effect of climate on processes and hence in controlling relief development.

nappe. A sheet-like rock unit, essentially the upper limb of a recumbent fold, that has been thrust many kilometres forward.

nivation. The combined effect of geomorphological processes operating beneath, and at the margin of, snow patches. These include transportation by abundant meltwater, due to frozen subsoil or permafrost limiting percolation, as well as frost action and solifluction.

nomothetic. A little-used adjective, adopted into geography's vocabulary to describe the approach in which cases and areas are studied as universals, with the formulation of general laws as the main objective.

Oligocene. A geological epoch of the Lower Tertiary period, following the Eocene and preceding the Miocene. It began some 36 million years ago and lasted some 13 million years, when the main phase of Alpine orogenesis began.

Ordovician. The second earliest geological period of the Paleozoic era, following the Cambrian and preceding the Silurian. It began some 500 million years or more ago, and lasted until about 430 million years ago.

orogeny. A mountain-building period or processes. Orogenesis involves metamorphism, plutonism, and plastic deformation in deeper layers as well as thrusting, faulting and folding in the lithosphere.

palaeo-, paleo-. A prefix meaning old. Thus composite terms like paleoclimate, paleoform, paleogeography, paleohydrology and paleorelief refer to conditions and products of the geological past.

Paleozoic. A geological era, following the Precambrian and preceding the Mesozoic. It lasted for some 350 million years, comprising the Cambrian, Ordovician, Silurian, Devonian, Carboniferous and Permian periods.

paradigm. A conceptual model showing previously unconnected ideas in a coherent pattern.

pedimentation. The combined effect of processes of lateral planation and transportation by streams, and the backwearing of the adjacent mountain front.

pediplain. A smooth, gently concave land surface in a desert region, attributed to the coalescence of adjacent pediments.

pedology. The scientific study of the soil.

pelagic deposits. Deep-sea sediments which have fallen from upper waters of the open ocean and are largely devoid of terrigenous material.

peneplanation. The combined effect of geomorphological processes of subaerial erosion in wearing down a structurally stable land surface to a low-lying plain of faint relief over an extended period of geological time.

periglacial. An adjective describing the climate, physical processes and associated soils and landforms in cold but ice-free environments. The seasonal change from winter freezing to summer thaw is the fundamental characteristic.

period. The common word used in a restricted sense to designate that geological-time interval which is a subdivision of an era but comprises two or more epochs.

planate. An adjective describing an erosionally levelled, flat land surface.

plankton. A collective term for all aquatic, microscopic organisms that float or drift. Fossil plankton are important indicators of stratigraphical horizons and associated paleotemperatures over a wide area.

Pleistocene. A geological epoch of the Quaternary period, following the Pliocene of the Tertiary and preceding the Holocene. Formerly, popularly known as the Ice Age.

Pliocene. A geological epoch of the Tertiary period, following the Miocene and preceding the Pleistocene, beginning some 5.5 million years ago and lasting some 3.7 million years.

pluton. A deep-seated intrusion of coarse-grained igneous rock, cooled and solidified at some depth below the land surface, generally granitic in composition and texture.

Precambrian. The enormous span of geological time before the beginning of the Paleozoic, preceding the Cambrian period which began some 600 million years ago. Precambrian rocks are usually highly metamorphosed.

progradation. The seaward building of a shoreline by nearshore deposition.

Quaternary. The latest geological period of the Cenozoic era, following the Tertiary. It consists of the Pleistocene and Holocene epochs and spans the last 2–3 million years.

regolith. The unconsolidated mantle of weathered debris which overlies solid bedrock. Only its upper, biologically modified horizon is recognized as 'soil'.

shield. A very large, old and rigid part of the earth's crust. Commonly, shields are exposures of Precambrian rocks with a very gently convex land surface which suggested the analogy with a warrior's shield.

slip plane. A planar slip surface at which mass movement occurs.

solifluction. Literally, soil-flow, but defined in terms of a periglacial mass movement in which meltwater of seasonal thaw, being unable to percolate downwards, due to frozen subsoils or permafrost, saturates the surface materials. Terracette microforms may develop in the waterlogged, flowing soil. The term appeared in 1906, following research in the Falkland Islands.

stillstand. A stationary sealevel stand in relation to that of the land.

stochastic process. A statistical process in which the dependent variable is random and, hence, any outcome is uncertain, predictable only in terms of probabilities.

stratigraphy. The arrangement of strata in their chronological sequence.

strike-slip fault. A fault in which slip movement is parallel to its strike.

subaerial. An adjective which describes geomorphological processes which exist at the ground surface, operating in the open air, as opposed to those of marine, subsurface, or subglacial environments.

subduction. The geophysical mechanism by which one lithospheric plate overrides another, thrusting it downward into the asthenosphere.

swale. A long, narrow trough between beach ridges, running approximately parallel to the coast.

tectonism. A general term for geophysical movements within the lithosphere.

tephra. Fragmental material thrown out by volcanic eruption or explosion, incorporating all pyroclastics like bombs, ash, pumice and cinders.

terrigenous deposits. Shallow-water marine sediments derived from land surfaces.

Tertiary. The first geological period of the Cenozoic era, following the Cretaceous and preceding the Quaternary. It began some 65 million years ago and lasted until 2 – 3 million years ago. It comprises five epochs, the Paleocene, Eocene, Oligocene, Miocene and Pliocene, with the Alpine orogeny reaching its maximum in the Miocene.

thufur. An Icelandic term for earth hummocks.

tombolo. A sand or gravel coastal bar connecting an island to the mainland or linking islands.

Triassic. The first geological period of the Mesozoic era, following the Permian of the Paleozoic and preceding the Jurassic, beginning about 240 million years ago and lasting some 40 million years.

tsunami. A sea wave produced by tectonic or volcanic disturbance of the ocean floor.

vulcanism. Mechanisms by which magma is extruded on to the earth's surface or intruded at shallow depths.

Bibliography

1 Scientific context of geomorphology

INTRODUCTION (p. 1)

BROWN, E. H. (1975) 'The content and relationships of physical geography', *Geog. Jour.*, 141, 35–40.

CHORLEY, R. J. (1971) 'The role and relations of physical geography', *Prog. Geog.*, 3, 87–109.

CLAYTON, K. M. (1971) 'Geomorphology–a study which spans the geology/geography interface', *Quart. Jour. Geol. Soc.*, 127, 471–6.

DURY, G. H. (1972) 'Some current trends in geomorphology', *Earth-Sci. Rev.*, 8, 45–72.

GOULD, P. R. (1973) 'The open geography curriculum', *in* R. J. Chorley (ed.) *Directions in Geography*, London, Methuen, 253–84.

JOHNSON, D. W. (1929) 'The geographic prospect', *Ann. Ass. Amer. Geogrs*, 19, 168–231.

KITTS, D. B. (1976) 'Certainty and uncertainty in geology', *Am. Jour. Sci.*, 276, 29–46.

MACKINDER, H. J. (1887) 'On the scope and methods of geography', *Proc. Roy. Geog. Soc.*, 9, 141–74.

WOOLDRIDGE, S. W. (1932) 'The Cycle of Erosion and the representation of relief', *Scot. Geog. Mag.*, 48, 30–6.

WOOLDRIDGE, S. W. (1958) 'The trend of geomorphology', *Trans. Inst. Brit. Geogrs*, 25, 29–35.

WORSLEY, P. (1979) 'Whither geomorphology?', *Area*, 11, 97–101.

IMPLICATIONS OF RECENT CHANGES OF EMPHASIS IN GEOLOGY (p. 2)

COTTON, C. A. (1960) *Geomorphology, an Introduction to the Study of Landforms*, Christchurch, Whitcombe & Tombs.

DREWRY, D. J. (1975) 'Initiation and growth of the East Antarctic ice sheet', *Quart. Jour. Geol. Soc.*, 131, 255–73.

GLAESER, J. D. (1978) 'Global distribution of barrier islands in terms of tectonic setting', *Jour. Geol.*, 86, 283–97.

GUEST, J. E. and MURRAY, J. B. (1979) 'An analysis of hazard from Mont Etna volcano', *Quart. Jour. Geol. Soc.*, 136, 347–54.

HENYEY, T. L. and LEE, T. C. (1976) 'Heat flow in Lake Tahoe, California-Nevada, and the Sierra Nevada – Basin and Range transition', *Bull. Geol. Soc. Am.*, 87, 1179–87.

HOFFMAN, J. I. (1979) 'The environmental geology curriculum', *Jour. Geol. Ed.*, 27, 114–16.

INMAN, D. L. and NORDSTROM, C. E. (1971) 'On the tectonic and morphologic classification of coasts', *Jour. Geol.*, 79, 1–21.

KELLER, J., RYAN, W. B. F., NINKOVICH, D. and ALTHERR, R. (1978) 'Explosive volcanic activity in the Mediterranean over the past 200,000 years as recorded in deep-sea sediments', *Bull. Geol. Soc. Am.*, 89, 591–604.

KNILL, J. (1970) 'Environmental geology', *Proc. Geol. Ass.*, 81, 529–37.

MITCHELL, A. H. and READING, H. G. (1971) 'Evolution of island arcs', *Jour. Geol.*, 79, 253–84.

NEWBERY, J. and SUBRAMANIAM, A. S. (1977) 'Geotechnical aspects of route-location studies for M4 north of Cardiff', *Quart. Jour. Eng. Geol.*, 10, 423–41.

NORMARK, W. R. and PIPER, D. J. W. (1972) 'Sediments and growth pattern of Navy deep-sea fan, San Clemente Basin, California Borderland', *Jour. Geol.*, 80, 198–223.

PITCHER, W. S. (1978) 'The anatomy of a batholith', *Quart. Jour. Geol. Soc.*, 135, 157–82.

POTTER, P. E. (1978) 'Significance and origin of big rivers', *Jour. Geol.*, 86, 13–33.

RICE, R. J. (1977) *Fundamentals of Geomorphology*, London, Longman.

ROMEY, W. D. (1972) 'Model for a humanistically oriented undergraduate geology department', *Jour. Geol. Ed.*, 20, 69–74.

SCHOLL, D. W., HEIN, J. R., MARLOW, M. and BUFFINGTON, E. C. (1977) 'Meiji sediment tongue: North Pacific evidence for limited movement between the Pacific and North American plates', *Bull. Geol. Soc. Am.*, 88, 1567–76.

SHEETS, P. D. and GRAYSON, D. K. (1979) *Volcanic Activity and Human Ecology*, New York, Academic Press.

UYEDA, S. and MIYASHIRO, A. (1974) 'Plate tectonics and the Japanese Islands: synthesis', *Bull. Geol. Soc. Am.*, 85, 1159–70.

WILSON, J. T. (1965) 'A new class of faults and their bearing on continental drift', *Nature*, 207, 343–7.

WOODLAND, A. W. (1968) 'Field geology and the civil engineer', *Proc. Yorks. Geol. Soc.*, 36, 531–78.

WYLLIE, P. J. (1976) *The Way the Earth Works: An Introduction to the New Global Geology and its Revolutionary Development*, New York, Wiley.

DIVERGENCE BETWEEN GEOMORPHOLOGY AND CERTAIN EMPHASES IN GEOLOGY (p. 6)

BOWEN, D. Q. (1973) 'The Pleistocene succession of the Irish Sea', *Proc. Geol. Ass.*, 84, 249–72.

BRIGGS, D. J. (1975) 'Origin, depositional environment and age of the Cheltenham Sand and Gravel and related deposits', *Proc. Geol. Ass.*, 86, 333–48.

BROWN, E. H. and WATERS, R. S. (1974) 'Geomorphology in the United Kingdom since the First World War', *in* E. H. Brown and R. S. Waters (eds) *Progress in Geomorphology*, Inst. Brit. Geogrs, Spec. Publ., 7, 3–9.

CLAYTON, K. M. (1980) 'Geomorphology', *in* E. H. Brown (ed.) *Geography Yesterday and Tomorrow*, London, Oxford University Press, 167–80.

GREEN, C. P. and MCGREGOR, D. F. M. (1980) 'Quaternary evolution of the River Thames', *in* D. K. C. Jones (ed.) *Shaping of Southern England*, London, Academic Press, Inst. Brit. Geogrs Spec. Publ., 11, 177–202.

KIDSON, C. and HEYWORTH, A. (1976) 'The Quaternary deposits of the Somerset Levels', *Quart. Jour. Eng. Geol.*, 9, 217–35.

ROSE, J., ALLEN, P. and HEY, R. W. (1976) 'Middle Pleistocene stratigraphy in southern East Anglia', *Nature*, 263, 492–4.

STRAW, A. (1973) 'The glacial geomorphology of central and north Norfolk', *East Midld Geogr.*, 5, 333–54.

WRIGHT, H. E. and FREY, D. G. (eds) (1965) *The Quaternary of the United States*, Princeton, NJ, Princeton University Press.

GEOGRAPHICAL CHARACTERISTICS OF GEOMORPHOLOGY (p. 8)

ABRAHAMS, A. D. (1977) 'The factor of relief in the evolution of channel networks in mature drainage basins', *Am. Jour. Sci.*, 277, 626–46.

BOND, G. (1978) 'Evidence for late-Tertiary uplift of Africa relative to North America, South America, Australia and Europe', *Jour. Geol.*, 86, 47–65.

CAILLEUX, A. and TRICART, J. (1956) 'Le problème de la classification des faits géomorphologiques', *Ann. Géogr.*, 65, 162–86.

CLAPPERTON, C. M. (1977) 'Volcanoes in space and time', *Prog. Phys. Geog.*, 1, 375–411.

GERRARD, A. J. W. and ROBINSON, D. A. (1971) 'Variability in slope measurements: discussion of the effects of different recording intervals and micro-relief in slope studies', *Trans. Inst. Brit. Geogrs*, 54, 45–54.

GUPTA, A. (1975) 'Stream characteristics in eastern Jamaica, an environment of seasonal flow and large floods', *Am. Jour. Sci.*, 275, 825–47.

INMAN, D. L. and NORDSTROM, C. E. (1971) 'On the tectonic and morphologic classification of coasts', *Jour. Geol.*, 79, 1–21.

LARSON, E. E., REYNOLDS, R. L., OZIMA, M., AOKI, Y., KINOSHITA, H., ZASSHU, S., KAWAI, N., NAKAJIMA, T., HIROOKA, K., MERRILL, R. and LEVI, S. (1975) 'Paleomagnetism of Miocene volcanic rocks of Guam and the curvature of the Southern Mariana Island arc', *Bull. Geol. Soc. Am.*, 86, 346–50.

MATSUDA, T., OTA, Y., ANDO, M. and YOKEKURA, N. (1978) 'Fault mechanism and recurrence time of major earthquakes in southern Kanto district, Japan, as deduced from coastal terrace data', *Bull. Geol. Soc. Am.*, 89, 1610–18.

MOOSER, F., MEYER-ABICH, H. and McBIRNEY, A. R. (1958) *Catalogue of Active Volcanoes: Central America*, Naples, International Association of Volcanology.

MORAN, S. R., CLAYTON, L, HOOKE, R. LeB., FENTON, M. M. and ANDRIASHEK, L. D. (1980) 'Glacier-bed landforms of the Prairie region of North America', *Jour. Glaciol.*, 25, 457–76.

PATTON, P. C. and SCHUMM, S. A. (1981) 'Ephemeral-stream processes: implications for studies of Quaternary valley fills', *Quat. Res.*, 15, 24–43.

PETERSON, J. A. (1968) 'Cirque morphology and Pleistocene ice-formation conditions in south-eastern Australia', *Aust. Geog. Studies*, 6, 67–83.

PITTY, A. F. (1971) *Introduction to Geomorphology*, London, Methuen.

ROSE, W. I., GRANT, N. K., HAHN, G. A., LANGE, I. M., POWELL, J. L., EASTER, J. and DEGRAFF, J. M. (1977) 'The evolution of Santa Maria volcano, Guatemala', *Jour. Geol.*, 85, 63–87.

SCHOLZ, C. H. (1977) 'Transform fault systems of California and New Zealand: similarities in their tectonic and seismic styles', *Quart. Jour. Geol. Soc.*, 133, 215–29.

TRENHAILE, A. S. (1976) 'Cirque morphometry in the Canadian Cordillera', *Ann. Ass. Amer. Geogrs*, 66, 451–62.

TRICART, J. (1965) *Principes et méthodes de la géomorphologie*, Paris, Masson.

WALKER, G. P. L. (1971) 'Grain-size characteristics of pyroclastic deposits', *Jour. Geol.*, 79, 696–714.

WALLING, D. E. and WEBB, B. W. (1978) 'Mapping solute loadings in an area of Devon, England', *Earth Surf. Processes*, 3, 85–99.

WILSON, L. (1973) 'Variations in mean annual sediment yield as a function of mean annual precipitation', *Am. Jour. Sci.*, 273, 335–49.

IMPLICATIONS OF CHANGES OF EMPHASIS IN GEOGRAPHY (p. 13)

BLANT, J. M. (1979) 'The dissenting tradition', *Ann. Ass. Amer. Geogrs*, 69, 157–64.

CARGO, D. N. and MALLORY, B. F. (1974) *Man and his Geologic Environment*, Reading, Mass., Addison-Wesley.

CARR, M. J. and STOIBER, R. E. (1977) 'Geologic setting of some destructive earthquakes in Central America', *Bull. Geol. Soc. Am.*, 88, 151–6.

CARRARA, A. and MERENDA, L. (1976) 'Landslide inventory in northern Calabria, southern Italy', *Bull. Geol. Soc. Am.*, 87, 1153–62.

COATES, D. R. (ed.) (1971) *Environmental Geomorphology*, Binghamton, State University of New York.

DAVIS, W. M. (1906) 'An inductive study of the content of geography', *Bull. Am. Geog. Soc.*, 38, 67–84.

HARTSHORNE, R. (1939) *The Nature of Geography: A Critical Survey of Current Thought in the Light of the Past*, Lancaster, PA, Association of American Geographers.

HARTSHORNE, R. (1959) *Perspectives on the Nature of Geography*, Chicago, Rand McNally.

HETTNER, A. (1972) *The Surface Features of the Land: Problems and Methods of Geomorphology*, London, Macmillan, translated by P. Tilley.

ILLIES, J. H. and GREINER, G. (1978) 'Rhinegraben and the Alpine system', *Bull. Geol. Soc. Am.*, 89, 770–82.

JOHNSON, D. W. (1929) 'The geographic prospect', *Ann. Ass. Amer. Geogrs*, 19, 168–231.

LEIGHLY, J. (1955) 'What has happened to Physical Geography?', *Ann. Ass. Amer. Geogrs*, 45, 309–18.

LOMNITZ, C. (1970) 'Casualties and behavior of populations during earthquakes', *Bull. Seismol. Soc. Am.*, 60, 1309–13.

MACKINDER, H. (1921) 'Geography as a pivotal subject in education', *Geog. Jour.*, 57, 376–84.

MACKINDER, H. (1931) 'The human habitat', *Scot. Geog. Mag.*, 47, 321–35.

MACKINKO, G. (1973) 'Man and the environment: a sampling of the literature', *Geog. Rev.*, 63, 378–91.

MIKESELL, M. W. (1969) 'The borderlands of geography as a social science', *in* M. Sherif and C. W. Sherif (eds) *Interdisciplinary Relationships in the Social Sciences*, Chicago, Aldine, 227–48.

MILLER, A. A. (1931) *Climatology*, London, Methuen.

PITTY, A. F. (1979) 'Conclusions', *in* A. F. Pitty (ed.) *Geographical Approaches to Fluvial Processes*, Norwich, Geo Books, 261–80.

SAUER, C. O. (1941) 'Foreword to historical geography', *Ann. Ass. Amer. Geogrs*, 31, 1–24.

SHEAFFER, J. R., ELLIS, D. W. and SPIEKER, A. M. (1969) 'Flood-hazard mapping in metropolitan Chicago', *US Geol. Surv. Circ. 601-C*.

SOMERVILLE, M. (1848) *Physical Geography*, London, John Murray, 7th rev. edn 1877.

TAAFFE, E. J. (1974) 'The spatial view in context', *Ann. Ass. Amer. Geogrs*, 64, 1–16.

TAAFFE, E. J. (1979) 'Geography of the Sixties in the Chicago area', *Ann. Ass. Amer. Geogrs*, 69, 133–8.

TUAN, Y-F. (1972) 'Environmental psychology: a review', *Geog. Rev.*, 62, 245–56.

WARD, R. C. (1967) *Principles of Hydrology*, London, McGraw-Hill, 2nd rev. edn 1975.

WOOLDRIDGE, S. W. and MORGAN, R. S. (1937) *The Physical Basis of Geography: An Outline of Geomorphology*, London, Longman.

YOUNG, K. (1975) *Geology: The Paradox of Earth and Man*, Boston, Houghton Mifflin.

STATISTICS AND GEOGRAPHY (p. 18)

BARNETT, V. and LEWIS, T. (1978) *Outliers in Statistical Data*, Chichester, Wiley.

BARROWS, H. H. (1923) 'Geography as human ecology', *Ann. Ass. Amer. Geogrs*, 13, 1–14.

CROWE, P. R. (1938) 'On progress in geography', *Scot. Geog. Mag.*, 54, 1–19.

DICKINSON, R. E. (1969) *The Makers of Modern Geography*, London, Routledge and Kegan Paul.

FISHER, R. A. (1926) 'The arrangement of field experiments', *Jour. Min. Agric.*, 33, 503–13.

FISHER, R. A. (1963) *Statistical Methods for Research Workers*, Edinburgh, Oliver and Boyd, 13th edn.

GALTON, F. (1857) 'The exploration of arid countries', *Proc. Roy. Geog. Soc.*, 2, 60–77.

HÄGERSTRAND, T. (1967) *Innovation Diffusion as a Spatial Process*, Chicago, University of Chicago Press, trans. by A. Pred.

LEIGHLY, J. (1937) 'Methodologic controversy in nineteenth-century German geography', *Ann. Ass. Amer. Geogrs*, 27, 125–41.

RUSSELL, E. J. (1966) *A History of Agricultural Science in Great Britain, 1620–1954*, London, Allen and Unwin.

SAUER, C. (1956) 'The education of a geographer', *Ann. Ass. Amer. Geogrs*, 46, 287–99.

TUAN, Y.-F. (1975) 'Place: an experiential perspective', *Geog. Rev.*, 65, 151–65.

WRIGHT, J. K. (1952) *Geography in the Making: The American Geographical Society, 1851-1951*, New York, American Geographical Society.

THE FOURTH DIMENSION (p. 21)

APPLETON, J. H. (1963) 'The efficacy of the Great Australian Divide as a barrier to railway communication', *Trans. Inst. Brit. Geogrs*, 33, 101–22.

BIRKELAND, P. W. (1974) *Pedology, Weathering and Geomorphological Research*, New York, Oxford University Press.

DIETZ, R. S. (1977) 'Plate tectonics: a revolution in geology and geophysics', *Tectonophysics*, 38, 1–6.

FINCH, V. C. (1939) 'Geographical science and social philosophy', *Ann. Ass. Amer. Geogrs*, 29, 1–28.

McLEAN, R. (1967) 'Plan shape and orientation of beaches along the east coast, South Island', *NZ Geogr.*, 23, 16–22.

CONCLUSION (p. 24)

ALVAREZ, W. (1973) 'Ancient course of the Tiber River near Rome: an introduction to the Middle Pleistocene volcanic stratigraphy of central Italy', *Bull. Geol. Soc. Am.*, 84, 749–58.

2 The nature of geomorphology

DESCRIPTION AND INTERPRETATION (p. 26)

BAGNOLD, R. A. (1941) *The Physics of Blown Sand and Desert Dunes*, London, Methuen, 3rd edn 1960.

BAKER, V. R. (1977) 'Stream-channel response to floods, with examples from central Texas', *Bull. Geol. Soc. Am.*, 88, 1057–71.

BARSCH, D. and LIEDTKE, H. (1980) 'Principles, scientific value and practical applicability of the geomorphological map of the Federal Republic of Germany at the scale of 1:25,000 (GMK 25) and 1:100,000 (GMK 100)', *Zeit. f. Geomorph.*, suppbd 36, 296–313.

BOWDEN, K. L. and WALLIS, J. R. (1964) 'Effect of stream-ordering technique on Horton's laws of drainage composition', *Bull. Geol. Soc. Am.*, 75, 767–74.

BRICE, J. C. (1974) 'Evolution of meander loops', *Bull. Geol. Soc. Am.*, 85, 581–6.

CALKIN, A. and CAILLEUX, A. (1962) 'A quantitative study of cavernous weathering (taffonis) and its application to glacial chronology in Victoria Valley, Antarctica', *Zeit. f. Geomorph.*, 6, 317–24.

CARTER, C. S. and CHORLEY, R. J. (1961) 'Early slope development in an expanding stream system', *Geol. Mag.*, 98, 117–30.

COOK, F. A. and RAICHE, V. G. (1962) 'Simple transverse nivation hollows at Resolute, NWT', *Geog. Bull.*, 18, 79–85.

COOKE, R. U. (1970) 'Morphometric analysis of pediments and associated landforms in the Western Mojave Desert, California', *Am. Jour. Sci.*, 269, 26–38.

CULLING, E. W. H. (1956) 'The longitundinal profiles of Chilterns streams', *Proc. Geol. Ass.*, 67, 314–45.

DEMEK, J. (ed.) (1972) *Manual of Detailed Geomorphological Mapping*, Prague, Academia.

DURY, G. H. (1972) 'A partial definition of the term *pediment* with field tests in humid-climate areas of southern England', *Trans. Inst. Brit. Geogrs*, 57, 139–52.

EPSTEIN, E. and GRANT, W. J. (1967) 'Soil losses and crust formation as related to some soil physical properties', *Proc. Soil Sci. Soc. Am.*, 31, 547–50.

FINKEL, H. J. (1959) 'The barchans of southern Peru', *Jour. Geol.*, 67, 614–47.

GAMS, I. (1978) 'The polje: the problem of definition', *Zeit, f. Geomorph.*, 22, 170–81.

GARDINER, V. (1978) 'Redundancy and spatial organization of drainage basin form indices: an empirical investigation of data from north-west Devon', *Trans. Inst. Brit. Georgrs*, NS, 3, 416–31.

GARNER, H. F. (1974) *The Origin of Landscapes*, New York, Oxford University Press.

HOMANS, G. C. (1967) *The Nature of Social Science*, New York, Harcourt, Brace and World.

HORTON, R. E. (1945) 'Erosional development of streams and their drainage basins: hydrophysical approach to quantitative morphology', *Bull. Geol. Soc. Am.*, 56, 275–370.

HOWARD, A. D. (1977) 'Effect of slope on the threshold of motion and its application to orientation of wind ripples', *Bull. Geol. Soc. Am.*, 88, 853–6.

JAMES, P. E. (1967) 'On the origin and persistence of error in geography', *Ann. Ass. Amer. Geogrs*, 57, 1–24.

JAMES, W. R. and KRUMBEIN, W. C. (1969) 'Frequency distribution of stream-link lengths', *Jour. Geol.*, 77, 544–65.

JARVIS, R. S. (1976) 'Stream orientation structures in drainage networks', *Jour. Geol.*, 84, 563–82.

JARVIS, R. S. (1977) 'Drainage network analysis', *Prog. Phys. Geog.*, 1, 271–295.

KING, P. B. and SCHUMM, S. A. (eds) (1980) *The Physical Geography (Geomorphology) of William Morris Davis*, Norwich, Geo Books.

LANGBEIN, W. B. and SCHUMM, S. A. (1958) 'Yield of sediment in relation to mean annual precipitation', *Trans. Amer. Geophys. Union*, 39. 1076–84.

LeGRAND, H. E. (1960) 'Metaphor in geomorphic expression', *Jour. Geol.*, 68, 576–9.

LEOPOLD, L. B., WOLMAN, M. G. and MILLER, J. P. (1964) *Fluvial Processes in Geomorphology*, San Francisco, Freeman.

LIVINGSTONE, D. N and HARRISON, R. T. (1981) 'Meaning through metaphor: analogy as epistemology', *Ann. Ass. Amer. Geogrs*, 71, 95–107.

LUBOWE, J. K. (1964) 'Stream-junction angles in the dendritic drainage pattern', *Am. Jour. Sci,*. 262, 325–39.

MANNERFELT, C. M. (1945) 'Nagra glacialmorfologiska formelement', *Geog. Annaler*, 27, 3–239.

MILLER, V. C. (1953) 'A quantitative geomorphic study of drainage-basin characteristics in the Clinch Mountain area, Virginia and Tennessee', Department of Geography, Columbia University, New York, *Tech. Rept.*, 3.

MORISAWA, M. (1958) 'Measurements of drainage-basin outline form', *Jour. Geol.*, 66, 587–90.

MOSLEY, M. P. and PARKER, R. S. (1972) 'Allometric growth: a useful concept in geomorphology?', *Bull. Geol. Soc. Am.*, 83, 3669–74.

NORMAN, J. W., LEIBOWITZ, T. H. and FOOKES, P. G. (1975) 'Factors affecting the detection of slope instability with air photographs in an area near Sevenoaks, Kent', *Quart. Jour. Eng. Geol.*, 8, 159–76.

OLLIER, C. C. (1967) 'Landform description without stage names', *Aust. Geog. Studies*, 5, 73–80.

ONGLEY, E. D. (1968) 'Toward a precise definition of drainage-basin axis', *Aust. Geog. Studies*, 6, 84–8.

PILGRIM, A. T. and CONACHER, A. J. (1974) 'Causes of earthflows in the southern Chittering Valley, Western Australia', *Aust. Geog. Studies*, 12, 38–56.

REED, B., GALVIN, C. J. and MILLER, J. P. (1962) 'Some aspects of drumlin geometry', *Am. Jour. Sci.*, 260, 200–10.

RICHARDS, K. S. (1976) 'Complex width-discharge relations in natural river sections', *Bull. Geol. Soc. Am.*, 87, 199–206.

SAVIGEAR, R. A. G. (1965) 'A technique of morphological mapping', *Ann. Ass. Amer. Geogrs*, 55, 514–38.

SHREVE, R. L. (1966) 'Statistical law of stream numbers', *Jour. Geol.*, 74, 17–37.

SHRODER, J. F. (1976) 'Mass movement on Nyika Plateau, Malawi', *Zeit. f. Geomorph.*, 20, 56–77.

SMALL, R. J. (1978) *The Study of Landforms*, Cambridge, Cambridge University Press, 2nd edn.

SONU, C. J. and VAN BEEK, J. L. (1971) 'Systematic beach changes on the Outer Banks, North Carolina', *Jour. Geol.*, 79, 416–25.

SPATE, O. H. K. and JENNINGS, J. N. (1972) 'Australian geography, 1951–1971', *Aust. Geog. Studies*, 10, 113–40.

STODDART, D. R. (1967) 'Organisms and ecosystems as geographical models', in R. J. Chorley and P. Haggett (eds) *Models in Geography*, London, Methuen, 511–48.

STRAHLER, A. N. (1952) 'Hypsometric (area-altitude) analysis of erosional topography', *Bull. Geol. Soc. Am.*, 63, 1117–42.

TRICART, J., RAYNAL, R. and BESANÇON, J. (1972) 'Cônes rocheux, pédiments, glacis', *Ann. Géogr.*, 81, 1–24.

TUAN, Y-F. (1979) 'Sight and pictures', *Geog. Rev.*, 69, 413–22.

WHEELER, D. A. (1979) 'The overall shape of longitudinal profiles of streams', in A. F. Pitty (ed.) *Geographical Approaches to Fluvial Processes*, Norwich, Geo Books, 241–60.

WHITAKER, C. R. (1979) 'The use of the term "pediment" and related terminology', *Zeit. f. Geomorph.*, 23, 427–39.

WOLDENBERG, M. J. (1966) 'Horton's laws justified in terms of allometric growth and steady state in open systems', *Bull. Geol. Soc. Am.*, 77, 431–4.

PROCESS AND FORM (p. 41)

AKAGI, Y. (1980) 'Relations between rock type and the slope form in the Sonoran Desert, Arizona', *Zeit. f. Geomorph.*, 24, 129–40.

ASHWELL, I. Y. (1975) 'Glacial and late glacial processes in Western Iceland', *Geog. Ann.*, 57A, 225–45.

BAGNOLD, R. A. (1941) *The Physics of Blown Sand and Desert Dunes*, London, Methuen, 3rd edn 1960.

BALLANTYNE, C. K. (1978) 'The hydrologic significance of nivation features in permafrost areas', *Geog. Ann.*, 60A, 51–4.

BIRD, E. C. F. and DENT, O. F. (1966) 'Shore platforms on the south coast of New South Wales', *Aust. Geogr.*, 10, 71–80.

BOULTON, G. S., MORRIS, E. M., ARMSTRONG, A. A. and THOMAS, A. (1979) 'Direct measurement of stress at the base of a glacier', *J. Glaciol.*, 22, 3–24.

CATT, J. A. and HODGSON, J. M. (1976) 'Soils and geomorphology of the Chalk in south-east England', *Earth Surf. Proc.*, 1, 181–93.

CHURCH, M., STOCK, R. F. and RYDER, J. M. (1979) 'Contemporary sedimentary environments on Baffin Island, NWT, Canada: debris-slope accumulations', *Arctic Alp. Res.*, 11, 371–402.

COOKE, R. U. and REEVES, R. W. (1972) 'Relations between debris size and the slope of mountain fronts and pediments in the Mojave Desert, California', *Zeit. f. Geomorph.*, 16, 76–82.

CROWTHER, J. (1979) 'Limestone solution on exposed rock outcrops in West

Malaysia', *in* A. F. Pitty (ed.) *Geographical Approaches to Fluvial Processes*, Norwich, Geo Books, 31–50.

DAY, D. G. (1978) 'Drainage density changes during rainfall', *Earth Surf. Proc.*, 3, 319–26.

DOORNKAMP, J. C. (1974) 'Tropical weathering and the ultra-microscopic characteristics of regolith quartz on Dartmoor', *Geog. Ann.*, 56A, 73–82.

DUNKERLEY, D. L. (1979) 'The morphology and development of Rillen-karren', *Zeit. f. Geomorph.*, 23, 332–48.

EDEN, M. J. and GREEN, C. P. (1971) 'Some aspects of granite weathering and tor formation on Dartmoor, England, *Geog. Ann.*, 53A, 92–9.

HALLAM, A. (1971) 'Mesozoic geology and the opening of the North Atlantic', *Jour. Geol.*, 79, 129–57.

HALLET, B., LORRAIN, R. and SOUCHEZ, R. (1978) 'The composition of basal ice from a glacier sliding over limestones', *Bull. Geol. Soc. Am.*, 89, 314–20.

HAZELHOFF, L., VAN HOOF, P., IMESON, A. C. and KWAAD, F. J. P. M. (1981) 'The exposure of forest soil to erosion by earthworms', *Earth Surf. Proc. Landforms*, 6, 235–50.

HIGGINS, C. G. (1980) 'Nips, notches and the solution of coastal limestones: an overview of the problem with examples from Greece', *Est. Coast. Mar. Sci.*, 10, 15–30.

HOOKE, J. M. (1979) 'An analysis of the processes of river-bank erosion', *J. Hydrol.*, 42, 39–62.

JENNINGS, J. N. and SWEETING, M. M. (1963) 'The limestone ranges of the Fitzroy Basin, Western Australia: a tropical semi-arid karst', *Bonner Geog. Abhandl.*, 32, 1–60.

KING, C. A. M. (1959) *Beaches and Coasts*, London, Arnold, 2nd edn 1972.

LEDGER, D. C., LOVELL, J. P. B. and CUTTLE, S. P. (1980) 'Rate of sedimentation in Kelly Reservoir, Strathclyde', *Scot. Jour. Geol.*, 16, 281–5.

MOORE, T. R. (1979) 'Rainfall erosivity in East Africa', *Geog. Ann.*, 61A, 147–56.

PEMBERTON, M. (1980) 'Earth hummocks at low elevations in the Vale of Eden, Cumbria', *Trans. Inst. Brit. Geogrs*, NS 5, 487–501.

PITTY, A. F. (1969) 'A scheme for hillslope analysis: 1. Initial considerations and calculations', University of Hull, *Occ. Papers in Geog.*, 9.

REID, I. (1979) 'Seasonal changes in microtopography and surface depression storage of arable soils', *in* G. E. Hollis (ed.) *Man's Impact on the Hydrological Cycle*, Norwich, Geo Books, 19–30.

ROBINSON, G. (1963) 'A consideration of the relations of geomorphology and geography', *Prof. Geogr.*, 15, 13–17.

SCHUMM, S. A. and LICHTY, R. W. (1965) 'Time, space and causality in geomorphology', *Am. Jour. Sci.*, 263, 110–19.

SMITH, B. J. (1978) 'The origin and geomorphic implications of cliff-foot recesses and tafoni on limestone hamadas in the north-west Sahara', *Zeit. f. Geomorph.*, 22, 21–43.

STERNBERG, R. W. and MARSDEN, M. A. H. (1979) 'Dynamics, sediment transport and morphology in a tide-dominated embayment', *Earth Surf. Proc.*, 4, 117–39.

SWAN, S. B. St, C. (1970) 'Relationships between regolith, lithology and slope in a humid tropical region: Johor, Malaya', *Trans. Inst. Brit. Geogrs*, 51, 189–200.

THORN, C. E. (1976) 'Quantitative evaluation of nivation in the Colorado Front Range', *Bull. Geol. Soc. Am.*, 87, 1169–78.

THORN, C. E. and HALL, K. (1980) 'Nivation: an arctic–alpine comparison and reappraisal', *J. Glaciol.*, 25, 109–23.

TRUDGILL, S. T., LAIDLAW, I. M. S. and SMART, P. L. (1980) Soil–water residence times and solute uptake on a dolomite bedrock – preliminary results', *Earth Surf. Proc.*, 5, 91–100.

VINCENT, P. J. and CLARKE, J. V. (1976) 'The terracette enigma – a review', *Biul. Peryglacjalny*, 25, 65–77.

WAYLEN, M. J. (1979) 'Chemical weathering in a drainage basin underlain by Old Red Sandstone', *Earth Surf. Proc.*, 4, 167–78.

WELCH, D. M. (1970) 'Substitution of space for time in a study of slope development', *Jour. Geol.*, 78, 234–9.

WHALLEY, W. B. (1976) *Properties of Materials and Geomorphological Explanation*, London, Oxford University Press.

WILLIAMS, M. A. J. (1978) 'Termites, soils and landscape equilibrium in the Northern Territory of Australia', *in* J. L. Davies and M. A. J. Williams (eds) *Landform Evaluation in Australasia*, Canberra, Australian National University Press, 128–41.

WRIGHT, L. D. and THOM, B. G. (1977) 'Coastal depositional landforms: a morphologic approach', *Prog. Phys. Geog.*, 1, 412–59.

ZAKRZEWSKA, B. (1967) 'Trends and methods in landform geography', *Ann. Ass. Amer. Geogrs*, 57, 128–65.

QUALITATIVE AND QUANTITATIVE ASPECTS (p. 49)

ANDREWS, J. T. and ESTABROOK, G. (1971) 'Applications of information and graph theory to multivariate geomorphological analysis', *Jour. Geol.*, 79, 207–21.

BAGNOLD, R. A. (1941) *The Physics of Blown Sand and Desert Dunes*, London, Methuen, 3rd edn 1960.

CARTER, C. S. and CHORLEY, R. J. (1961) 'Early slope development in an expanding stream system', *Geol. Mag.*, 98, 117–30.

CHORLEY, R. J., DUNN, A. J. and BECKINSALE, R. P. (1964) *The History of the Study of Landforms*, Vol. 1, London, Methuen.

COOPER, R. G. (1980) 'A sequence of landsliding mechanisms in the Hambleton Hills, Northern England, illustrated by features at Peak Scar, Hawnsby', *Geog. Ann.*, 62A, 149–56.

138   *The Nature of Geomorphology*

DAVIES, G. L. (1969) *The Earth in Decay: A History of British Geomorphology, 1578–1878*, London, Macdonald.

DEVDARIANI, A. (1967) 'The profile of equilibrium and a regular regime', *Soviet Geog.*, 8, 168–83.

EPSTEIN, E. and GRANT, W. J. (1967) 'Soil losses and crust formation as related to some soil physical properties', *Proc. Soil Sci. Soc. Am.*, 31, 547–50.

EYLES, R. J. (1974) 'Bifurcation ratio – a useless index?' *NZ Geogr.*, 30, 166–71.

EYLES, R. J. (1977) 'Birchams Creek: the transition from a chain of ponds to a gully', *Aust. Geog. Studies*, 15, 146–57.

GILBERT, G. K. (1886) 'The inculcation of scientific method by example', *Am. Jour. Sci.*, 31, 284–99.

GOUDIE, A. S. (ed.) (1981) *Geomorphological Techniques*, London, Allen & Unwin.

GROVE, J. M. (1972) 'The incidence of landslides, avalanches and floods in western Norway during the Little Ice Age', *Alp. Arctic Res.*, 4, 131–8.

HICKIN, E. J. and NANSON, G. C. (1975) 'The character of channel migration on the Beatton river, north-east British Columbia, Canada', *Bull. Geol. Soc. Am.*, 86, 487–94.

LANGBEIN, W. B. and SCHUMM, S. A. (1958) 'Yield of sediment in relation to mean annual precipitation', *Trans. Am. Geophys. Union*, 39, 1076–84.

LeGRAND, H. E. (1962) 'Perspective on problems of hydrogeology', *Bull. Geol. Soc. Am.*, 73, 1147–52.

LEOPOLD, L. B. and LANGBEIN, W. B. (1962) 'The concept of entropy in landscape evolution', *US Geol. Surv. Prof. Paper 500–A*.

OSTERKAMP, W. R. (1978) 'Gradient, discharge and particle-size relations of alluvial channels in Kansas, with observations on braiding', *Am. Jour. Sci.*, 278, 1253–68.

PITTY, A. F. (1971) *Introduction to Geomorphology*, London, Methuen.

RICHARDS, K. S. (1977) 'Channel and flow geometry: a geomorphological perspective', *Prog. Phys. Geog.*, 1, 65–102.

SCHEIDEGGER, A. E. (1961) *Theoretical Geomorphology*, Berlin, Springer.

SCHEIDEGGER, A. E. (1979) 'Orientationsstruktur der Talanlagen in der Schweiz', *Geog. Helvetica*, 34, 9–16.

TAMBURI, A. J. (1974) 'Creep of single rocks on bedrock', *Bull. Geol. Soc. Am.*, 85, 351–6.

TROEH, F. R. (1965) 'Landform equations fitted to contour maps', *Am. Jour. Sci.*, 263, 616–27.

TWIDALE, C. R. (1978) 'Early explanation of granite boulders', *Rev. Géomorph. Dyn.*, 27, 133–42.

WEISS, D. (1974) 'Late Pleistocene stratigraphy and paleoecology of the Lower Hudson River estuary', *Bull. Geol. Soc. Am.*, 85, 1561–70.

WERRITY, A. and FERGUSON, R. I. (1980) 'Pattern changes in a Scottish braided river over 1, 30, and 200 years', *in* R. A. Cullingford, D. A.

Davidson and J. Lewin (eds), *Timescales in Geomorphology*, Chichester, Wiley, 53–68.

LABORATORY AND FIELD WORK (p. 56)

AGER, D. V. (1970) 'On seeing the most rocks', *Proc. Geol. Ass.*, 81, 421–7.

ARNETT, R. R. (1972) 'The field measurement of lateral soil-water movement', *Rev. Géomorph. Dyn.*, 21, 177–81.

BUNGE, W. W. (1973) 'The Geography', *Prof. Geogr.*, 25, 331–7.

DAY, M. J., LEIGH, C. and YOUNG, A. (1980) 'Weathering of rock discs in temperate and tropical soils', *Zeit. f. Geomorph.*, suppbd 35, 11–15.

DURY, G. H. (1976) 'Experimental study of river incision: discussion and reply', *Bull. Geol. Soc. Am.*, 87, 319–20.

GLEASON, R., BLACKLEY, M. W. L. and CARR, A. P. (1975) 'Beach stability and particle-size distribution, Start Bay', *Quart. Jour. Geol. Soc.*, 131, 83–101.

GRAF, W. L., TRIMBLE, S. W., TOY, T. J. and COSTA, J. E. (1980) 'Geographic geomorphology in the 'eighties', *Prof. Geogr.*, 32, 279–84.

HARRIS, C. D. (1979) 'Geography at Chicago in the 1930s and 1940s', *Ann. Ass. Amer. Geogrs*, 69, 21–32.

HOWARD, A. D. (1977) 'Effect of slope on the threshold of motion and its application to orientation of wind ripples', *Bull. Geol. Soc. Am.*, 88, 853–6.

IMESON, A. C., VIS, R. and de WATER, E. (1981) 'The measurement of water-drop impact forces with a piezo-electric transducer', *Catena*, 8, 83–96.

JAMES, P. E. and MATHER, C. (1977) 'The role of periodic field conferences in the development of geographical ideas in the United States', *Geog. Rev.*, 67, 446–61.

PARK, C. C. (1978) 'Channel-bank material and cone penetrometer studies: an empirical evaluation', *Area*, 10, 227–30.

PETTIJOHN, F. J. (1956) 'In defense of outdoor geology', *Bull. Amer. Ass. Petrol. Geols*, 40, 1455–61.

PITTY, A. F. (1979) 'Conclusions', *in* A. F. Pitty (ed.) *Geographical Approaches to Fluvial Processes*, Norwich, Geo Books, 261–80.

SHEPHERD, R. G. and SCHUMM, S. A. (1974) 'Experimental study of river incision', *Bull. Geol. Soc. Am.*, 85, 257–68.

STEINKER, D. C. (1979) 'The undergraduate field program as an essential element in geological education', *Jour. Geol. Ed.*, 27, 162–4.

SUNAMURA, T. (1975) 'A laboratory study of wave-cut platform formation', *Jour. Geol.*, 83, 389–97.

THORNES, J. B. and BRUNSDEN, D. (1977) *Geomorphology and Time*, London, Methuen.

WHITNEY, M. I. (1978) 'The role of vorticity in developing lineation by wind erosion', *Bull. Geol. Soc. Am.*, 89, 1–18.

WILLIAMS, A. T., GRANT, C. J. and LEATHERMAN, S. P. (1977) 'Sedimentation patterns in Repulse Bay, Hong Kong', *Proc. Geol. Ass.*, 88, 183–200.

THE ROLE OF GEOMORPHOLOGY (p. 61)

BOWLER, J. M. (1976) 'Aridity in Australia: age, origins and expression in aeolian landforms and sediments', *Earth-Sci. Rev.*, 12, 279–310.
BRUNSDEN, D., DOORNKAMP, J. C., FOOKES, P. G., JONES, D. K. C. and KELLY, M. H. (1975) 'Large-scale geomorphological mapping and highway engineering design', *Quart. Jour. Eng. Geol.*, 8, 227–322.
BRYAN, K. (1950) 'The place of geomorphology in the geographic sciences', *Ann. Ass. Amer. Geogrs*, 40, 196–208.
BURTON, I. and KATES, R. W. (1964) 'The floodplain and the seashore: a comparative analysis of hazard-zone occupance', *Geog. Rev.*, 54, 366–85.
COATES, D. R. (ed.) (1973) *Environmental Geomorphology and Landscape Conservation*, Stroudsburg, PA, Dowden, Hutchinson and Ross, 3 vols.
CONACHER, A. J. and DALRYMPLE, J. B. (1977) 'The nine-unit landsurface model: an approach to pedogeomorphic research', *Geoderma*, 18, 1–154.
COOKE, R. U. and DOORNKAMP, J. C. (1974) *Geomorphology in Environmental Management: An Introduction*, Oxford, Clarendon Press.
DERBYSHIRE, E. and EVANS, I. S. (1976) 'The climatic factor in cirque variation', *in* E. Derbyshire (ed.) *Geomorphology and Climate*, Chichester, Wiley, 447–94.
DOLAN, R. (1972) 'Barrier dune systems along the outer banks of North Carolina: a reappraisal', *Science*, 176, 286–8.
EYLES, R. J., CROZIER, M. J. and WHEELER, R. H. (1978) 'Landslips in Wellington City', *NZ Geogr.*, 34, 54–74.
FOSTER, H. D. (1976) 'Assessing disaster magnitude: a social-science approach', *Prof. Geogr*, 28, 241–7.
FRENCH, H. M. (1976) *The Periglacial Environment*, London, Longman.
GERASIMOV, I. P. (1976) 'Problems of natural environment transformation in Soviet constructive geography', *Prog. Geog.*, 9, 73–99.
GOULD, P. (1979) 'Geography 1957–1977: the Augean period', *Ann. Ass. Amer. Geogrs*, 69, 139–51.
HAILS, J. R. (ed.) (1977) *Applied Geomorphology*, Amsterdam, Elsevier.
IVES, J. D., MEARS, A. I., CARRARA, P. E. and BOVIS, M. J. (1976) 'Natural hazards in mountain Colorado', *Ann. Ass. Amer. Geogrs*, 66, 129–44.
JAKUCS, L. (1978) 'Physical-geographical and geological aspects of the exploration of the hydrocarbon reserves of the South Hungarian Plain', *Acta Geographica*, 18, 91–105.
JOHNSON, R. H. (1980) 'Hillslope stability and landslide hazard – a case study from Longdendale, north Derbyshire, England', *Proc. Geol. Ass.*, 91, 315–25.

KENNEDY, W. Q. (1962) 'Theoretical factors in geomorphological analysis', *Geol. Mag.*, 99, 304–12.

KERSHAW, A. P. (1976) 'A Late Pleistocene and Holocene pollen diagram from Lynch's Crater, north-eastern Queensland, Australia,' *New Phytol.*, 77, 469–98.

KREIGER, M. H. (1973) 'What's wrong with plastic trees?', *Science*, 179, 446–55.

MARKER, M. E. (1972) 'Karst landform analysis as evidence for climatic change in the Transvaal, South Africa', *S. Afr. Geogr.*, 54, 152–62.

MITCHELL, C. W. (1973) *Terrain Evaluation*, London, Longman.

OLLIER, C. D. (1977) 'Terrain classification – methods, applications and principles', *in* J. R. Hails (ed.) *Applied Geomorphology*, Amsterdam, Elsevier, 277–316.

O'RIORDAN, T. (1971) 'Environmental management', *Prog. Geog.*, 3, 173–231.

ORME, A. R. (1980) 'Energy-sediment interaction around a groin', *Zeit. f. Geomorph.*, suppbd 34, 111–28.

PARKES, J., PARKES, G. M. and DAY, J. C. (1975) 'The hazard of sensitive clays: a case study of the Ottawa-Hull area', *Geog. Rev.*, 65, 198–213.

PRICE, R. J. (1976) *Highland Landforms*, Edinburgh, Morison & Gibbs.

STREET, F. A. and GROVE, A. T. (1979) 'Global maps of lake-level fluctuations since 30,000 yr BP', *Quat. Res.*, 12, 83–118.

TORRY, W. I. (1979) 'Hazards, hazes and holes: a critique of *The Environment as Hazard* and general reflections on disaster research', *Can. Geog.*, 23, 368–83.

UNWIN, D. J. (1973) 'The distribution and orientation of corries in northern Snowdonia, Wales', *Trans. Inst. Brit. Geogrs*, 58, 85–97.

ZIEMNICKI, S. and REPELEWSKA-PEKOLOWA, J. (1972) 'Investigations into present-day geomorphological processes in the loess areas of the Lublin Plateau', *Geog. Polonica*, 23, 63–76.

3 Basic postulates

CATASTROPHISM AND UNIFORMITARIANISM (p. 66)

BAKER, J. N. L. (1948) 'Mary Somerville and geography in England', *Geog. Jour.*, 111, 207–22.

BAKER, V. R. (1977) 'Stream-channel response to floods, with examples from central Texas', *Bull. Geol. Soc. Am.*, 88, 1057–71.

BERGHINZ, C. (1971) 'Venice is sinking into the sea', *Civ. Eng.*, 41, 67–71.

BJÖRNSSON, H. (1975) 'Explanation of jökulhlaups from Grímsvötn, Vatnajökull, Iceland', *Jökull* (Ár), 24, 1–26.

BROWN, J. A. F. (1972) 'Hydrologic effects of a bushfire in a catchment in south-eastern New South Wales', *J. Hydrol.*, 15, 77–96.

BURKE, K. C. A. and DEWEY, J. F. (1973) 'An outline of Precambrian plate development', *in* D. H. Tarling and S. K. Runcorn (eds) *Continental Drift, Sea-floor Spreading and Plate Tectonics*, New York, Academic Press.

CUNNINGHAM, F. F. (1977) 'Lyell and uniformitarianism', *Can. Geogr.*, 21, 164–74.

DANSGAARD, W., JOHNSON, S. J., CLAUSEN, H. B. and LANDWAY, C. C. (1972) 'Speculations about the next glaciation', *Quat. Res.*, 2, 396–8.

DAVIES, D. K., QUEARRY, M. W. and BONIS, S. B. (1978) 'Glowing avalanches from the 1974 eruption of the volcano Fuego, Guatemala', *Bull. Geol. Soc. Am.*, 89, 369–84.

DINGLE, R. V. (1977) 'The anatomy of a large submarine slump on a sheared continental margin (SE Africa)', *Quart. Jour. Geol. Soc.*, 134, 293–310.

DONN, W. L. and SHAW, D. M. (1977) 'Model of climate evolution based on continental drift and polar wandering', *Bull. Geol. Soc. Am.*, 88, 390–6.

DOUGLAS, I. (1967) 'Man, vegetation and sediment yield of rivers', *Nature*, 215, 925–8.

EMBLETON, C. and KING, C. A. M. (1968) *Glacial and Periglacial Geomorphology*, London, Arnold.

FEININGER, T. (1971) 'Chemical weathering and glacial erosion of crystalline rocks and the origin of till', *US Geol. Surv. Prof. Paper 750-C*, 65–81.

FLOHN, H. (1979) 'On time scales and causes of abrupt paleoclimatic events', *Quat. Res.*, 12, 135–49.

GALTON, F. (1863) 'The avalanches of the Jungfrau', *Alpine Jour.*, 1, 184–8.

HEY, R., JOHNSON, G. L. and LOWRIE, A. (1977) 'Recent plate motions in the Galapagos area', *Bull. Geol. Soc. Am.*, 88, 1385–1403.

HOPLEY, D. (1974) 'The cyclone Althea storm surge', *Aust. Geog. Studies*, 12, 90–106.

La CHAPELLE, E. R and LAND, T. E. (1980) 'A comparison of observed and calculated avalanche velocities', *Jour. Glaciol.*, 25, 309–14.

LAINE, E. P. (1980) 'New evidence from beneath the western North Atlantic for the depth of glacial erosion in Greenland and North America', *Quat. Res.*, 14, 188–98.

LÜDER, H-J. (1980) 'Zur Abschätzung der Abflusskapazität eines Wadis der nördlichen Sahara (Djebel es Soda/Libyen): Bericht über Gelände-untersuchungen im November und Dezember 1977', *Die Erde*, 111, 121–32.

MARCUS, M. G. (1960) 'Periodic drainage of glacier-dammed Tulsequah Lake, British Columbia', *Geog. Rev.*, 50, 89–106.

NYE, J. F. (1976) 'Water flow in glaciers: jökulhlaups, tunnels and veins', *Jour. Glaciol.*, 17, 181–207.

OLLIER, C. D. (1979) 'Evolutionary geomorphology of Australia and Papua New Guinea', *Trans. Inst. Brit. Geogrs*, NS 4, 516–39.

OSTERKAMP, W. R. (1978) 'Gradient, discharge, and particle-size relations of alluvial channels in Kansas, with observations on braiding', *Am. Jour. Sci.*, 278, 1253–68.

POTTER, P. E. (1978a) 'Significance and origin of big rivers', *Jour. Geol.*, 86, 13–33.

POTTER, P. E. (1978b) 'Petrology and chemistry of modern big-river sands', *Jour. Geol.*, 86, 423–49.

SOMERVILLE, M. (1848) *Physical Geography*, London, J. Murray, 7th edn 1877.

TERNAN, J. L. and WILLIAMS, A. G. (1979) 'Hydrological pathways and granite weathering on Dartmoor', *in* A. F. Pitty (ed.) *Geographical Approaches to Fluvial Processes*, Norwich, Geo Books, 5–30.

WHALLEY, W. B. (1971) 'Observations of the drainage of an ice-dammed lake – Strupvatnet, Troms, Norway', *Norsk Geog. Tidsskr.*, 25, 165–74.

YOSHIKAWA, T. (1974) 'Denudation and tectonic movement in contemporary Japan', *Bull. Dept. Geog. Univ. Tokyo*, 6, 1–14.

THE CYCLE OF EROSION (p. 74)

BANDOIAN, C. A. and MURRAY, R. C. (1974) 'Pliocene-Pleistocene carbonate rocks of Bonaire, Netherlands Antilles', *Bull. Geol. Soc. Am.*, 85, 1243–52.

DALZELL, D. and DURRANCE, E. M. (1980) 'The evolution of the Valley of Rocks, North Devon', *Trans. Inst. Brit. Geogrs*, NS 5, 66–79.

FENNEMAN, N. M. (1936) 'Cyclic and non-cyclic aspects of erosion', *Bull. Geol. Soc. Am.*, 47, 173–86.

FLINN, D. (1977) 'The erosion history of Shetland: a review', *Proc. Geol. Ass.*, 88, 129–46.

GIRDLER, R. W. (1965) 'The formation of new oceanic crust', *Phil. Trans. Roy. Soc.*, A, 258, 123–36.

JONES, D. K. C. (1981) *Southeast and Southern England*, London, Methuen.

KING, L. C. (1976) 'Planation remnants upon high lands', *Zeit. f. Geomorph.*, 20, 133–48.

KÖNIG, M. and TALWANI, M. (1977). 'A geophysical study of the southern continental margin of Australia: Great Australian Bight and western sections', *Bull. Geol. Soc. Am.*, 88, 1000–14.

MCCONNELL, R. B. (1968) 'Planation surfaces in Guyana', *Geog. Jour.*, 134, 506–20.

MEISLER, H. (1962) 'Origin of erosional surfaces in the Lebanon Valley, Pennsylvania', *Bull. Geol. Soc. Am.*, 73, 1071–82.

MENARD, H. W. (1960) 'The East Pacific rise', *Science*, 132, 1737–46.

MOUSINHO de MEIS, M. R. and MONTEIRO, A. M. F. (1979) 'Upper Quaternary "rampas": Doce river valley, south-eastern Brazilian plateau', *Zeit. f. Geomorph.*, 23, 132–51.

NOBLE, D. C., MCKEE, E. H. and MEGARD, F. (1978) 'Eocene uplift and unroofing of the coastal batholith near Lima, central Peru', *Jour. Geol.*, 86, 403–5.

PALMER, M. V. and PALMER, A. N. (1975) 'Landform development in the Mitchell Plain of southern Indiana: origin of a partially karsted plain', *Zeit. f. Geomorph.*, 19, 1–39.

PITTY, A. F. (1965) 'A study of some escarpment gaps in the southern Pennines', *Trans. Inst. Brit. Geogrs*, 37, 127–45.

PITTY, A. F. (1968) 'The scale and significance of solutional loss from the limestone tract of the southern Pennines', *Proc. Geol. Ass.*, 79, 153–77.

PROFFETT, J. M. (1977) 'Cenozoic geology of the Yerington district, Nevada, and implications for the nature and origin of Basin and Range faulting', *Bull. Geol. Soc. Am.*, 88, 247–66.

RUTLAND, R. W. R. (1973) 'On the interpretation of Cordilleran orogenic belts', *Am. Jour. Sci.*, 273, 811–49.

SCHUMM, S. A. (1963) 'The disparity between present rates of denudation and orogeny', *US Geol. Surv. Prof. Paper 454-H*, 1–13.

THOMPSON, G. A. and BURKE, D. B. (1973) 'Rate and direction of spreading in Dixie Valley, Basin and Range Province, Nevada', *Bull. Geol. Soc. Am.*, 84, 627–32.

TILLEY, P. (1968) 'Early challenges to Davis's concept of the Cycle of Erosion', *Prof. Geogr.*, 20, 265–9.

WHEELER, D. A. (1979) 'The overall shape of longitudinal profiles of streams', *in* A. F. Pitty (ed.) *Geographical Approaches to Fluvial Processes*, Norwich, Geo Books, 241–60.

CLIMATIC GEOMORPHOLOGY (p. 82)

BENEDICT, J. B. (1976) 'Frost creep and gelifluction features: a review', *Quat. Res.*, 6, 55–76.

BIROT, P. (1966) *General Physical Geography*, London, Harrap, trans. by M. Ledésert.

BROOK, G. A. and FORD, D. C. (1978) 'The origin of labyrinth and tower karst and the climatic conditions necessary for their development', *Nature*, 275, 493–6.

BÜDEL, J. (1963) 'Klima-genetische Geomorphologie', *Geol. Rundschau*, 15, 269–85.

BÜDEL, J. (1969) 'Das System der klimagenetischen Geomorphologie', *Erdkunde*, 23, 165–83.

CAINE, N. (1978) 'Climatic geomorphology in mid-latitude mountains', *in* J. L. Davies and M. A. J. Williams (eds) *Landform Evolution in Australasia*, Canberra, Australian National University Press, 113–27.

CLAYTON, K. M. (1971) 'Geomorphology – a study which spans the geology/ geography interface', *Quart. Jour. Geol. Soc.*, 127, 471–6.

COTTON, C. A. (1942) *Climatic Accidents in Landscape-making, A Sequel to 'Landscape as developed by the processes of normal erosion'*, Christchurch, Whitcombe and Tombs.

DERBYSHIRE, E. (ed.) (1973) *Climatic Geomorphology*, London, Macmillan.

DERBYSHIRE, E. (ed.) (1976) *Geomorphology and Climate*, Chichester, Wiley.

DURY, G. H. (1972) 'A partial definition of the term *pediment* with field tests in humid-climate areas of southern England', *Trans. Inst. Brit. Geogrs*, 57, 139–52.

GREGORY, K. J. and GARDINER, V. (1975) 'Drainage density and climate', *Zeit. f. Geomorph.*, 19, 287–98.

HARRIS, S. A. (1979) 'Ice caves and permafrost zones in south-west Alberta', *Erdkunde*, 33, 61–70.

HAYES, M. O. (1967) 'Relationship between coastal climate and bottom sediment type on the inner continental shelf', *Marine Geol.*, 5, 111–32.

HOLMES, G. W., HOPKINS, D. M. and FOSTER, H. L. (1968) 'Pingos in central Alaska', *US Geol. Surv. Bull. 1241-H*.

HOLZNER, L. and WEAVER, G. D. (1965) 'Geographical evaluation of climatic and climato-genetic geomorphology', *Ann. Ass. Amer. Geogrs*, 55, 592–602.

HORTON, R. E. (1945) 'Erosional development of streams and their drainage basins: hydro-physical approach to quantitative morphology', *Bull. Geol. Soc. Am.*, 56, 275–370.

JOHN, B. S. and SUGDEN, D. (1975) 'Coastal geomorphology of high latitudes', *Prog. Geog.*, 7, 53–132.

KENNEDY, B. A. (1976) 'Valley-side slopes and climate', *in* E. Derbyshire (ed.) *Geomorphology and Climate*, Chichester, Wiley, 171–201.

KING, L. C. (1953) 'Canons of landscape evolution', *Bull. Geol. Soc. Am.*, 64, 721–62.

KING, L. C. (1957) 'The uniformitarian nature of hillslopes', *Trans. Edin. Geol. Soc.*, 17, 81–102.

LEOPOLD, L. B., WOLMAN, M. G. and MILLER, J. P. (1964) *Fluvial Processes in Geomorphology*, San Francisco, Freeman.

MENSCHING, H. (1970) 'Flächenbildung in der Sudan und Sahel Zone (Ober-Volta und Niger)', *Zeit. f. Geomorph.*, suppbd 10, 1–29.

OBERLANDER, T. M. (1972) 'Morphogenesis of granite boulder slopes in the Mojave desert, California', *Jour. Geol.*, 80, 1–20.

PISSART, A. (1965) 'Les pingos des Hautes Fanges: les problèmes de leur genèse', *Ann. Soc. Géol. Belg.*, 88, 277–89.

PITTY, A. F. (1979) *Geography and Soil Properties*, London, Methuen.

POTTER, P. E. (1978) 'Petrology and chemistry of modern big-river sands', *Jour. Geol.*, 86, 423–49.

SELBY, M. J. (1971) 'Slopes and their development in an ice-free arid area of Antarctica', *Geog. Ann.*, 53A, 235–45.

SHORT, A. D. (1975) 'Offshore bars along the Alaskan coast', *Jour. Geol.*, 83, 209–21.

SINGER, A. (1980) 'The paleoclimatic interpretation of clay minerals in soils and weathering profiles', *Earth-Sci. Rev.*, 15, 303–26.

SPARKS, B. W., WILLIAMS, R. B. G. and BELL, F. G. (1972) 'Presumed ground-ice depressions in East Anglia', *Proc. Roy. Soc. Lond.*, A, 327, 329–43.

STODDART, D. R. (1969) 'Climatic geomorphology: a review and re-assessment', *Prog. Geog.*, 1, 159–222.

SWEETING, M. M. (1972) *Karst Landforms*, London, Macmillan.

THOMAS, M. F. (1974) *Tropical Geomorphology*, London, Macmillan.

TOY, T. J. (1977) 'Hillslope form and climate', *Bull. Geol. Soc. Am.*, 88, 16–22.

TRICART, J. and CAILLEUX, A. (1972) *Introduction to Climatic Geomorphology*, London, Longmans, trans. by C. J. Kiewiet de Jonge.

VANN, J. H. (1980) 'Shoreline changes in mangrove areas', *Zeit. f. Geomorph.*, suppbd 34, 255–61.

WATSON, E. (1971) 'Remains of pingos in Wales and the Isle of Man', *Geol. Jour.*, 7, 381–92.

WILHELMY, H. (1958) *Klimamorphologie der Massengesteine*, Braunschweig, Westermann.

WILKINSON, T. J. and BUNTING, B. T. (1975) 'Overland transport of sediment by rill water in a periglacial environment in the Canadian High Arctic', *Geog. Ann.*, 57A, 105–16.

STILLSTANDS AND THE MOBILITY OF EARTH STRUCTURES (p. 93)

ANDREWS, J. T. (1978) 'Sealevel history of arctic coasts during the Upper Quaternary: dating, sedimentary sequences and history', *Prog. Phys. Geog.*, 2, 377–407.

ÅSE, L-E. (1980) 'Shore displacement at Stockholm during the last 1000 years', *Geog. Ann.*, 62A, 83–91.

AXELROD, D. I. and BAILEY, H. P. (1968) 'Paleotemperature analysis of Tertiary floras', *Palaeogeog. Palaeoclim. Palaeoecol.*, 6, 163–95.

BLOOM, A. L. (1967) 'Pleistocene shorelines: a new test of isostasy', *Bull. Geol. Soc. Am.*, 78, 1477–94.

BOND, G. (1978) 'Evidence for late Tertiary uplift of Africa, relative to North America, South America, Australia and Europe', *Jour. Geol.*, 86, 47–65.

BOWMAN, I. (1926) 'The analysis of landforms', *Geog. Rev.*, 16, 122–32.

BRADLEY, W. C. and GRIGGS, G. B. (1976) 'Form, genesis, and deformation of central California wave-cut platforms', *Bull. Geol. Soc. Am.*, 87, 433–49.

FARRAND, W. R. (1962) 'Post-glacial uplift in North America', *Am. Jour. Sci.*, 260, 181–99.

FOX, P. J., HEEZEN, B. C. and JOHNSON, G. L. (1970) 'Jurassic sandstone from the tropical Atlantic', *Science*, 170, 1402–4.

GROSSWALD, M. G. (1980) 'Late Weichselian Ice Sheet of northern Eurasia', *Quat. Res.*, 13, 1–32.

HALLAM, A. (1971) 'Re-evaluation of the palaeogeographic argument for an expanding earth', *Nature*, 232, 180–2.

HANCOCK, J. M. and KAUFFMANN, E. G. (1979) 'The great transgressions of the late cretaceous', *Quart. Jour. Geol. Soc.*, 136, 175–86.

KVASOV, D. D. and VERBITSKY, M. Ya (1981) 'Causes of Antarctic glaciation in the Cenozoic', *Quat. Res.*, 15, 1–17.

MACHIDA, H., NAKAGAWA, H. and PIRAZZOLI, P. (1976) 'Preliminary study of the Holocene sealevels in the central Ryukyn Islands', *Rev. Géomorph Dyn.*, 25, 49–62.

MORISAWA, M. (1975) Tectonics and geomorphic models', *in* W. N. Melhorn and R. C. Flemal (eds) *Theories of Landform Development*, New York, Binghamton, 199–216.

PITMAN, W. C. (1978) 'Relationship between eustacy and stratigraphic sequences of passive margins', *Bull. Geol. Soc. Am.*, 89, 1389–1403.

RAO, M. S. and VAIDYANADHAN, R. (1979) 'Morphology and evolution of Godavari delta, India', *Zeit. F. Geomorph.*, 23, 243–55.

RONA, P. A. (1973) 'Relation between rates of sediment accumulation on continental shelves, sea-floor spreading and eustacy inferred from the central North Atlantic', *Bull. Geol. Soc. Am.*, 84, 2851–72.

ROSSI, G. (1980) 'Tectonique, surfaces d'aplanissement et problèmes de drainage au Rwanda-Burundi', *Rev. géomorph. dyn.*, 29, 81–100.

SUNAMURA, T. (1978) 'A model of the development of continental shelves having erosional origin', *Bull. Geol. Soc. Am.*, 89, 504–10.

WATTS, A. B. and RYAN, W. B. F. (1976) 'Flexure of the lithosphere and continental margin basins', *Tectonophysics*, 36, 25–44.

WOODS, A. J. (1980) 'Geomorphology, deformation and chronology of marine terraces along the Pacific coast of Central Baja California, Mexico', *Quat. Res.*, 13, 346–64.

YOSHIKAWA, T. (1974) 'Denudation and tectonic movement in contemporary Japan', *Bull. Dept. Geog. Univ. Tokyo.*, 6, 1–14.

STRUCTURE, PROCESS AND STAGE (p. 100)

BANDOIAN, C. A. and MURRAY, R. C. (1974) 'Pliocene-Pleistocene carbonate rocks of Bonaire, Netherlands Antilles', *Bull. Geol. Soc. Am.*, 85, 1243–52.

BEAUMONT, P. (1978) 'Man's impact on river systems: a world-wide view', *Area*, 10, 38–41.

BORÓWKA, R. K. (1980) 'Present-day dune processes and dune morphology on the Łeba barrier, Polish coast of the Baltic', *Geog. Ann.*, 62A, 75–82.

BRAY, J. R. (1974) 'Volcanism and glaciation during the last 40 millenia', *Nature*, 252, 679–80.

CALKIN, P. E. and BRETT, C. E. (1978) 'Ancestral Niagara River drainage: stratigraphic and paleontologic setting', *Bull. Geol. Soc. Am.*, 89, 1140–54.

COLHOUN, E. A. and SYNGE, F. M. (1980) 'The cirque moraines at

Lough Nahanagan, County Wicklow, Ireland', *Proc. Roy. Irish Acad.*, 80B, 25–45.

DARBY, H. C. (1940) *The Draining of the Fens*, Cambridge, Cambridge University Press.

DAVIES, G. L. H. and STEPHENS, N. (1978) *Ireland*, London, Methuen.

DUBOIS, R. N. (1978) 'Beach topography and beach cusps', *Bull. Geol. Soc. Am.*, 89, 1133–9.

FOLEY, M. G. (1978) 'Scour and fill in steep, sand-bed ephemeral streams', *Bull. Geol. Soc. Am.*, 89, 559–70.

GODARD, A. (1979) 'Reconnaissance dans l'extrémité Nord du Labrador et du NouveauQuébec: contribution à l'étude géomorphologique des socles des milieux froids', *Rev. Géomorph. Dyn.*, 28, 125–42.

GRAF, W. L. (1976) 'Stream, slopes and suburban development', *Geog. Analysis*, 8, 153–73.

GUPTA, A. (1975) 'Stream characteristics in eastern Jamaica, an environment of seasonal flow and large floods', *Am. Jour. Sci.*, 275, 825–47.

HACK, J. T. (1960) 'Interpretation of erosional topography in humid temperate regions', *Am. Jour. Sci.*, 258A (Bradley volume), 80–97.

HACK, J. T. (1965) 'Geomorphology of the Shenandoah Valley, Virginia and West Virginia, and origin of the residual ore deposits', *US Geol. Surv. Prof. Paper 484.*

HAYS, J. D., IMBRIE, J. and SHACKLETON, N. J. (1976) 'Variations in the earth's orbit: pacemaker of the ice ages', *Science*, 194, 1121–32.

HAYS, J. D. and PITMAN, W. C. (1973) 'Lithospheric plate motion, sealevel changes and climatic and ecological consequences', *Nature*, 246, 18–21.

HELGREN, D. M. (1979) 'Relict channelways of the middle Orange river', *S. Afr. Jour. Sci.*, 75, 462–3.

HILLS, E. S. (1961) 'Morphotectonics and the geomorphological sciences, with special reference to Australia', *Quart. Jour. Geol. Soc.*, 117, 77–89.

INMAN, D. L. and NORDSTROM, C. E. (1971) 'On the tectonic and morphologic classification of coasts', *Jour. Geol.*, 79, 1–21.

KANEPS, A. (1979) 'Gulf Stream velocity fluctuations during the late Cenozoic', *Science*, 204, 297–301.

KNIGHTON, A. D. (1975) 'Variations in at-a-station hydraulic geometry', *Am. Jour. Sci.*, 275, 186–218.

KNUDSEN, N. T. and THEAKSTONE, W. H. (1981) 'Recent changes of the glacier Østerdalsisen, Svartisen, Norway', *Geog. Ann.*, 63A, 23–30.

KOCH, A. J. and McLEAN, H. (1975) 'Pleistocene tephra and ash-flow deposits in the volcanic highlands of Guatemala', *Bull. Geol. Soc. Am.*, 86, 529–41.

LYELL, C. (1830–3) *Principles of Geology, Being an Attempt to Explain the Former Changes of the Earth's Surface, by Reference to Causes Now in Operation*, London, Murray.

MALEY, J. (1977) 'Palaeoclimates of central Sahara during the Holocene', *Nature*, 269, 573–7.

MARGOLIS, S. V. and KENNETT, J. P. (1971) 'Cenozoic paleoglacial history of Antarctic recorded in sub-Antarctic deep-sea cores', *Am. Jour. Sci.*, 271, 1–36.

MENARD, H. W. (1961) 'Some rates of regional erosion', *Jour. Geol.*, 69, 154–61.

MOTTERSHEAD, D. N. and WHITE, I. D. (1972) 'The lichenometric dating of glacier recession, Tunsbergdal, southern Norway', *Geog. Ann.*, 54A, 47–52.

NEWSON, M. (1980) 'The erosion of drainage ditches and its effect on bed-load yields in mid-Wales: reconnaissance case studies', *Earth Surf. Proc.*, 5, 275–90.

OLLIER, C. D. and PAIN, C. F. (1978) 'Geomorphology and tectonics of Woodlark Island, Papua New Guinea', *Zeit. f. Geomorph.*, 22, 1–20.

ONGLEY, E. D. (1974) 'Fluvial morphometry on the Cobar pediplain', *Ann. Ass. Amer. Geogrs*, 64, 281–92.

OTA, Y. and NOGAMI, M. (1979) 'Recent research in Japanese geomorphology', *Prof. Geog.*, 31, 410–16.

PARK, C. C. (1981) 'Man, river systems and environmental impacts', *Prog. Phys. Geog.*, 5, 1–31.

PETTS, G. (1979) 'Complex response of river-channel morphology subsequent to reservoir construction', *Prog. Phys. Geog.*, 3, 329–62.

RAPP, A., MURRAY-RUST, D. H., CHRISTIANSSON, C. and BERRY, L. (1972) 'Soil erosion and sedimentation in four catchments near Dodoma, Tanzania', *Geog. Ann.*, 54A, 255–318.

RASID, H. (1979) 'The effects of regime regulation by the Gardiner dam on downstream geomorphic processes in the South Saskatchewan River', *Can. Geog.*, 23, 140–58.

SANDFORD, K. S. (1954) 'River development and superficial deposits', *in* A. F. Martin and R. W. Steel (eds) *The Oxford Region*, London, Oxford University Press, 21–4.

SCHNITKER, D. (1980) 'Global paleoceanography and its deep-water linkage to the Antarctic glaciation', *Earth-Sci. Rev.*, 16, 1–20.

SISSONS, J. B. (1974) 'The Quaternary in Scotland: a review', *Scot. Jour. Geol.*, 10, 311–37.

SWIFT, D. J. P., STANLEY, D. J. and CURRAY, J. R. (1971) 'Relict sediments on continental shelves: a reconsideration', *Jour. Geol.*, 79, 322–46.

TWIDALE, C. R. (1976) 'On the survival of paleoforms', *Am. Jour. Sci.*, 276, 77–95.

WARWICK, G. T. (1976) 'Geomorphology and caves', *in* T. D. Ford and C. H. D. Cullingford (eds) *The Science of Speleology*, London, Academic Press, 61–125.

WEERTMAN, J. (1976) 'Milankovitch solar radiation variations and ice-age ice-sheet sizes', *Nature*, 261, 17–20.

WOOD, P. A. (1976) 'Stream characteristics in eastern Jamaica, an environment of seasonal flow and large floods', *Am. Jour. Sci.*, 276, 917–18.

YOUNG, R. A. and BRENNAN, W. J. (1974) 'Peach Springs Tuff: its bearing on structural evolution of the Colorado Plateau and development of Cenozoic drainage in Mohave County, Arizona', *Bull. Geol. Soc. Am.*, 85, 83–90.

THE NECESSITY FOR SIMPLIFICATION OF GEOMORPHOLOGICAL COM-
PLEXITY (p. 110)

BAGNOLD, R. A. (1941) *The Physics of Blown Sand and Desert Dunes*, London, Methuen, 3rd edn 1960.
BLONG, R. J. (1975) 'Hillslope morphometry and classification: a New Zealand example', *Zeit. f. Geomorph.*, 19, 405–29.
BRADLEY, W. C. and GRIGGS, G. B. (1976) 'Form, genesis and deformation of central California wave-cut platforms', *Bull. Geol. Soc. Am.*, 87, 433–49.
CHORLEY, R. J. and BENNETT, R. J. (1978) *Environmental Systems: Philosophy, Analysis and Control*, London, Methuen.
CHORLEY, R. J. and KENNEDY, B. (1971) *Physical Geography: A Systems Approach*, London, Prentice-Hall.
DAVIES, D. K., QUEARRY, M. W. and BONIS, S. B. (1978) 'Glowing avalanches from the 1974 eruption of the volcano Fuego, Guatemala', *Bull. Geol. Soc. Am.*, 89, 369–84.
ELLENBERG, L. (1978) 'Coastal types of Venezuela – an application of coastal classifications', *Zeit. f. Geomorph.*, 22, 439–56.
FERGUSON, R. I. (1973) 'Sinuosity of supraglacial streams', *Bull. Geol. Soc. Am.*, 84, 251–6.
FISK, H. N. (1977) 'Magnitude and frequency of transport of solids by streams in the Mississippi basin', *Am. Jour. Sci.*, 277, 862–75.
FLOYD, B. and O'BRIEN, D. (1976) 'Whither geography? A cautionary tale from economics', *Area*, 8, 15–23.
FREEMAN, T. W. (1961) *A Hundred Years of Geography*, London, Duckworth.
GILBERT, G. K. (1914) 'The transportation of debris by running water', *US Geol. Surv. Prof. Paper 86*.
GRAF, W. L. (1977) 'The rate law in fluvial geomorphology', *Am. Jour. Sci.*, 277, 178–91.
HOOKE, R. LeB. (1975) 'Distribution of sediment transport and shear stress in a meander bend', *Jour. Geol.*, 83, 543–65.
HOWARD, A. D. (1970) 'Simulation of stream networks by headward growth and branching', *Geog. Analysis*, 3, 29–50.
HUNTLEY, D. A. and BOWEN, A. J. (1975) 'Field observations of edge waves and their effect on beach material', *Quart. Jour. Geol. Soc.*, 131, 69–81.
JAMES, W. R. and KRUMBEIN, W. C. (1969) 'Frequency distribution of stream-link lengths', *Jour. Geol.*, 77, 544–65.
KENNEDY, B. A. (1979) 'A naughty world', *Trans. Inst. Brit. Geogrs*, NS 4, 550–8.

KING, C. A. M. (1970) 'Feedback relationships in geomorphology', *Geog. Ann.*, 52A, 147–59.

LANGBEIN, W. B. and LEOPOLD, L. B. (1964) 'Quasi-equilibrium states in channel morphology', *Am. Jour. Sci.*, 262, 782–94.

LUDWICK, J. C. (1974) 'Tidal currents and zigzag sand shoals in a wide estuary entrance', *Bull. Geol. Soc. Am.*, 85, 717–26.

OLLIER, C. D. (1968) 'Open systems and dynamic equilibrium in geomorphology', *Aust. Geog. Studies*, 6, 167–70.

PARSONS, J. J. (1977) 'Geography as exploration and discovery', *Ann. Ass. Amer. Geogrs*, 67, 1–16.

PINCHEMEL, P. (1954) *Les plaines de craie du nord-ouest du Bassin Parisien et du sud-est du Bassin de Londres et leurs bordures*, Paris, Armand Colin.

PITTY, A. F. (1966) 'Some problems in the location and delimitation of slope Pennines', *Trans. Inst. Brit. Geogrs*, 37, 127–45.

PITTY, A. F. (1966) 'Some problems in the location and delimination of slope profiles', *Zeit. f. Geomorph.*, 10, 454–61.

ROBINSON, K. W. (1970) 'Diversity, conflict and change – the meeting place of geography and politics', *Aust. Geog. Studies*, 8, 1–15.

SCHUMM, S. A. (1956) 'Evolution of drainage systems and slopes in badlands at Perth Amboy, New Jersey', *Bull. Geol. Soc. Am.*, 67, 597–646.

SHORT, A. D. and WRIGHT, L. D. (1974) 'Lineaments and coastal geomorphic patterns in the Alaskan Arctic', *Bull. Geol. Soc. Am.*, 85, 931–6.

SMITH, N. D. (1974) 'Sedimentology and bar formation in the Upper Kicking Horse River, a braided outwash stream', *Jour. Geol.*, 82, 205–23.

SPARKS, B. W. (1971) *Rocks and Relief*, London, Longman.

STRAHLER, A. H. (1950) 'Equilibrium theory of erosional slopes approached by frequency distribution analysis', *Am. Jour. Sci.*, 248, 673–96; 800–14.

TUTTLE, S. D. (1975) 'How many peneplains can sit on the top of a mountain?', *in* W. N. Melhorn and R. C. Flemal (eds) *Theories of Landform Development*, New York, Binghamton, 299–306.

WATKINS, N. D. and WALKER, G. P. L. (1977) 'Magnetostratigraphy of eastern Iceland', *Am. Jour. Sci.*, 277, 513–84.

Index

mountain, belts, 8; building, 123; front, 78–9, 123; ranges, 2, 121
mountains, 72, 77, 87, 102
mud, 71, 72, 84, 86

nappes, 8, 123
natural, activity, 69; beauty, 55; cliffs, 59; conditions, 60; control, 116–17; disasters, 16, 56; environment, 58–60, 121; eustatic gauge, 116; flows, 68; geometrical simplification, 117; hazards, 6, 16, 65, 67, 71; phenomena, 111; processes, 58, 62, 66, 74, 113; relationships, 58; resources, 16; science, 15, 18–20, 40, 49, 62, 107; situation, 60; streams, 60; surroundings, 65
nivation, 41, 42, 123; hollow, 12, 33, 37, 38, 41
nomothetic approach, 45, 123
non-cyclic, erosion, 108; explanations, 78; terms, 93
non-monotonic relationships, 52, 53
notch, coastal, 48; marine-cut, 75

ocean, 67, 70; basins, 10, 12, 98, 103, 120; circulations, 103; currents, 98, 117; floor, 10, 125; -floor cores, 4, 5; trenches, 2; volume, 99; waves, 72, 125
oceans, 87, 88, 103, 123
Oligocene, 88, 98, 103, 121, 123, 125
Ollier, C. D., 70
organic matter, 42, 46; decomposition, 22; horizons, 88
organisms, 27, 67, 103, 119, 121, 124
orientation, 9, 34, 37; corrie, 62; joint, 55; ripple, 60; valley, 55
orogenesis, 8; Alpine, 123, 125
orogenic, belts, 4, 49; pulse, 5
outcrop, 19, 45, 54, 61, 79, 80, 93, 110, 122
outlier, geological, 19; statistical, 19, 51
outwash plains, 116
overbank stage, 28–9

paleoclimate, 62, 87–9, 123
paleoforms, 87–9, 90
paleogeography, 12, 24, 123; dynamic, 67
paleorelief, 77, 123
paleosols, 106
paleotemperatures, 88, 124
Paleotiber, 24–5
paroxysms, 67, 71, 72
particle, movement, 9; shape, 41, 50; sorting, 10–11
particle size, 10–11, 41, 51, 52, 59, 92, 115; analysis, 58
particles, 9, 13, 44–5, 49, 59, 73
Passarge, S., 81
patterned ground, 30
Pearson, Karl, 18

peat, 62
pebbles, 13, 58, 106
pedimentation, 77, 78, 123
pediments, 9, 26, 27, 34, 45, 46–8, 53, 88, 91, 123
pediplains, 79, 123
Penck, A., 83
Penck, W., 45, 93
peneplain, 27, 75, 77–8, 81, 96, 101
peneplanation, 75, 79, 93, 123
periglacial, 123, 124; action, 73, 88–9; environment, 119; episodes, 88; features, 116; landsurfaces, 41; microforms, 43; regime, 79; zone, 83, 87, 90
period, geological, 124
permafrost, 30, 42, 84–5, 88–91, 117, 123, 124
permeability, 58
photographs, 28, 30; air, 30
physical, principles, 82–3, 92; properties, 101
physics, 40, 48–9, 54, 55, 91, 108
pingos, 43, 88–9, 116
plain, erosional, 75, 78, 124
planation, 79; lateral, 123
plane table, 30
plankton, 124
planning, 15, 65; flood-plain, 16; land-use, 16; proposals, 64
plate, Asian, 10; boundaries, 2, 4, 8, 115; Cococ, 74; continental, 2, 4, 10, 17; converging, 2; lithospheric, 3, 9, 119, 125; margins, 67, 74, 96; micro-, 13; Nazca, 74; North American, 3; oceanic, 2, 10–11; overridden, 2; Pacific, 3, 74; tectonics, 2–4, 99, 102, 120
platforms, 95, 99–100
Pleistocene, 8, 24, 71–2, 89, 95, 96, 99, 100, 116, 122, 124; deposits, 7; history, 49; sealevel fluctuations, 8; succession, 7
Pliocene, 24, 69, 95, 103, 104, 105, 124
Plio-Pleistocene, 98
ploughing, 72
pluton, 5, 41, 124
polar wandering, 4, 69
polje, 26
pollen, 62, 72
polygons, 12, 90, 91
pool-riffle spacing, 53
population, 16–17; biology, 16
postdiction, 67, 72–3
precipitation, 10, 52, 53, 86, 87; effective, 52
predictions, 54, 55, 67, 74, 113
processes, 1, 2, 9, 24, 27, 28–9, 41–6, 50, 54, 55, 57, 58, 62, 66, 72–5, 82, 83, 87, 88, 93, 95, 100–10, 123; azonal, 90; biochemical, 58; biological, 58, 82; chemical, 82; coastal, 45; depositional, 11, 102; endogenetic, 72, 121; erosional, 10, 59, 102; fluvial, 45; geological, 41; geophysical, 2; mechanical,

Milton Keynes UK
Ingram Content Group UK Ltd.
UKHW040053071024
449327UK00019B/519

9 780367 224158